Studies in Applied Philosophy, Epistemology and Rational Ethics

Volume 12

Editor-in-Chief

Lorenzo Magnani, University of Pavia, Italy
lmagnani@unipv.it

About This Series

Studies in Applied Philosophy, Epistemology and Rational Ethics (SAPERE) publishes new developments and advances in all the fields of philosophy, epistemology, and ethics, bringing them together with a cluster of scientific disciplines and technological outcomes: from computer science to life sciences, from economics, law, and education to engineering, logic, and mathematics, from medicine to physics, human sciences, and politics. It aims at covering all the challenging philosophical and ethical themes of contemporary society, making them appropriately applicable to contemporary theoretical, methodological, and practical problems, impasses, controversies, and conflicts. The series includes monographs, lecture notes, selected contributions from specialized conferences and workshops as well as selected PhD theses.

Editorial Board

For further volumes:
http://www.springer.com/series/10087

Nir Fresco

Physical Computation and Cognitive Science

 Springer

Nir Fresco
School of Humanities & Languages,
    University of New South Wales
Sydney
Australia

ISSN 2192-6255   ISSN 2192-6263   (electronic)
ISBN 978-3-662-50943-2   ISBN 978-3-642-41375-9   (eBook)
DOI 10.1007/978-3-642-41375-9
Springer Heidelberg New York Dordrecht London

Printed on acid-free paper

Springer is part of Springer Science+Business Media (www.springer.com)

To my beloved wife, Calanit, and my lovely daughters, Leah & Noa

Foreword

Nir Fresco tackles the nontrivial question of "What is computation?". His thorough review includes explication and critique of accounts of computation ranging from Turing's, as is embodied by the Turing machine, to more recent accounts from the computational theory of mind. He offers his own account of concrete digital computation that is based on a critical analysis of the role information plays in computation and how differing notions of information fare in the context of computation. His two-pronged approach will appeal to philosophically inclined computer scientists who want to understand better common theoretical claims in cognitive science. This book provides a thorough and timely analysis of differing accounts of computation while advancing the important role that information plays in understanding computation.

--Marty J. Wolf, Professor of Computer Science,
Bemidji State University

Preface

It is my great pleasure to write this preface for Dr. Nir Fresco's book "Physical Computation and Cognitive Science" based on his doctoral thesis "Concrete Digital Computation: Competing Accounts and its Role in Cognitive Science" of which I was the supervisor. Nir was enrolled in Philosophy at UNSW as a part-time PhD student while working full-time as a computing professional, a very reflective one. Not only did he submit his thesis early he also found time to publish six papers related to it on the way through. A truly remarkable achievement.

Nir's main project in this book is to clarify the nature of computation, but he has a special interest in its explanatory role in cognitive science and understanding cognition. Computation is an ambiguous concept and computer scientists, philosophers and cognitive scientists who use the concept can contest some claim using it and not realise they are not actually in disagreement with each other, even though it looks as though they are. This cuts both ways: sometimes they talk past each other and think they are in agreement when they are not. A practical aim of this book is to help with this problem by clarifying the nature of computation.

In very brief compass but sufficient to point to the substantial scope of this volume, here is a selective description of its eight chapters.

The first sets the stage and discusses computation in cognitive science and a distinction between three kinds of representation in computation: intrinsic, mathematical and extrinsic. The central question to be answered is what does it take for a system to perform physical digital computation and some important distinctions start the discussion here.

Chapters 2 through to 7 form the heart of the book's answer to this question. Chapter 2, propaedeutic in nature, discusses adequacy criteria for evaluating accounts of computation. The criteria proposed by five influential theorists are considered: Brian Cantwell Smith, Gualtiero Piccinini, John von Neumann, David Chalmers and Matthias Scheutz. Three of Smith's criteria are discussed, six of Piccinini's and three of von Neumann's as well as the single criterion proposed by Chalmers and Scheutz. There is some overlap among these and Nir argues cogently for six criteria broadly in agreement with Piccinini.

These six criteria concern what an adequate account should explain and classify. They are, using Nir's terminology, the following: the conceptual, implementation, dichotomy, miscomputation, taxonomy and program execution criteria. The first requires being able to explain basic concepts in computation such as that of a computer program, algorithm and computer architecture. The second concerns the implementation by physical systems of abstract computation. The third requires the accurate classification of paradigmatic computing systems distinguishing them from non-computing systems. The fourth and fifth require accounting for miscomputation and appropriately classifying computational systems of differing powers, respectively. The sixth requires an account to explain the relation between program execution and concrete digital computation.

Adequacy criteria having been chosen, we are in a position to see how well competing accounts of computation measure up. Chapter 3, as Nir says in its title, starts at the beginning and examines Alan Turing's account. Unsurprisingly at this stage, but informatively his account does not do well. Turing was answering a related but different question.

Chapter 4 may be seen as a kind of interlude in the proceedings, not really discussing a serious account of computation. Titled, 'The Triviality "Account" Examined', it discusses attempts by two philosophers John Searle and Hilary Putnam to undermine the computational view of cognition by arguing in effect that every physical object (of sufficient everyday complexity) computes, trivialising the notion of what it is to compute. The exercise of arguing against this view is instructive.

The next three chapters deal successively with semantic accounts of computation in Chapter 5, information processing accounts in Chapter 6 and more heterogeneously causal and functional accounts in Chapter 7.

None of the three semantic accounts in Chapter 5 passes muster as accounts of concrete computation. On these accounts of computation, representation is the central notion and, more specifically, extrinsic (roughly, external world) representation. These semantic accounts are the physical symbol system account (PSS), the formal symbol manipulation account (FSM) and a reconstruction of B.C. Smith's participatory account. FSM has been the most influential of these accounts especially in cognitive science and the critique here should contribute to a useful rethink.

Two accounts do better, one each from chapters 6 and 7. In Chapter 6 the Instructional Information Processing account (IIP) does well as does the Mechanistic account in Chapter 7. Neither of these appeals to extrinsic representation for the explanation of computation.

Chapter 6 on computation as information processing explores what this common view could mean. Nir distinguishes four kinds of information: two non-semantic kinds (Shannon and algorithmic) and two semantic kinds (declarative and instructional). He argues that only an instructional information processing account, a view he introduces and develops here, can satisfy the adequacy criteria.

Chapter 7 focuses on three fairly recent non-semantic accounts of concrete digital computation. The first two of these are primarily causal and the last causal and functional. The first, called the Gandy-Sieg account, synthesises the older views of Robin Gandy and the more recent work of Wilfried Sieg. Concentrating on machine computation this view requires the system's operation to be describable as a sequence of state transitions and has an emphasis on physical requirements for computation and, hence, (on some interpretations) implementation.

The second of these accounts is the algorithm execution account. As the name suggests this view ties computation to acting in accordance with an algorithm. There is an older view of Robert Cummins that is significantly refined by Jack Copeland so as not to succumb to trivialisation (as in Chapter 3). To this end Copeland puts constraints on implementation, requiring a specification of the basic architecture of the system and an algorithm specific to the primitive operations of the architecture as well as the possibility of a labelling of the system's parts subject to special conditions.

The third view is Piccinini's mechanistic account. This non-semantic not exclusively causal view focuses on the functional/organisational features of the physical computing system. A central notion is that of a 'digit', a stable discrete state of a component of the system whose type can be reliably distinguished by it. Concatenations of digits are called 'strings' and can be either data or rules depending on their functional role in the system. On this view, a digital computing system is a mechanism with the function of mapping input strings (paired perhaps with internal states) into output strings according to general rules.

This quick review of these important accounts omits the instructive discussions of the way they fail or pass the six criteria of adequacy.

The subtlety of thought on show in the first seven chapters bears further fruit in the last when applied, among other things, to the role of computation in cognitive science. Chapter 8 is titled 'Computation Revisited in the Context of Cognitive Science' and covers a lot of ground. Two early positions defended in the first two sections are firstly that nontrivial computations typically only process implicit intrinsic and mathematical representations, not extrinsic ones and secondly that computational explanations of cognition will be unintelligible unless there is commitment to a single interpretation of the key phrase 'digital computation'. The largest third section on the explanatory role of computation in cognitive science discusses the explanatory frameworks of computationalism, connectionism and dynamicism arguing that they are not mutually exclusive and opening the way for their integration. This section also discusses computational neuroscience, neural computation and the nature of mechanistic and non-mechanistic explanations. The chapter closes with some general remarks on computation and cognition.

A final remark about this book and its provenance and author. In the second half of 2013 Nir was awarded the international Goldberg Memorial Prize. This prize, awarded by the International Association for Computing and Philosophy, is for outstanding graduate research. As noted above, Nir already had 6 publications written during his PhD candidacy. Ideas from all of these contributed to his doctorate and provided some of the rationale for this prestigious award. They will find a wider audience here.

Dr. Phillip Staines
Philosophy Discipline
School of Humanities and Languages
University of New South Wales

Abstract

There are currently considerable confusion and disarray about just how we should view computationalism, connectionism and dynamicism as explanatory frameworks in cognitive science. A key source of this ongoing conflict among the central paradigms in cognitive science is an equivocation on the notion of computation *simpliciter*. 'Computation' is construed differently by computationalism, connectionism, dynamicism and computational neuroscience. I claim that these central paradigms, properly understood, can contribute to an integrative cognitive science. Yet, before this claim can be defended, a better understanding of 'computation' is required.

'Digital computation' is an ambiguous concept. It is not just the classical dichotomy between analogue and digital computation that is the basis for the equivocation on 'computation' *simpliciter* in cognitive science, but also the diversity of extant accounts of *digital* computation. What does it take for a physical system to perform digital computation? There are many answers to this question ranging from Turing machine computation, through the formal manipulation of symbols, the execution of algorithms and others, to strong pancomputationalism, according to which every physical system computes every Turing-computable function. Despite some overlap among them, extant accounts of concrete digital computation are intensionally and extensionally non-equivalent, thereby rendering 'digital computation' ambiguous.

The objective of this book is twofold. First, it is to promote a clearer understanding of concrete digital computation. Accordingly, the main underlying thesis of the book is that not only are extant accounts of concrete digital computation non-equivalent, but most of them are inadequate. In the course of examining several key accounts of concrete digital computation, I also propose the instructional information processing account, according to which nontrivial digital computation is the processing of discrete data in accordance with finite instructional information. The second objective is to establish the foundational role of computation in cognitive science whilst rejecting the extrinsically representational nature of computation proper.

Keywords: Cognitive Science, Computability, Computationalism, Concrete Computation, Connectionism, Data, Digital Computation, Information, Turing Machine.

Acknowledgements

A preliminary version of this book was submitted as a PhD thesis in May 2012 at the School of Humanities & Languages, Faculty of Arts & Social Sciences, University of New South Wales, Sydney, Australia.

My greatest debt is to Phillip Staines, my PhD advisor, who has worked closely with me all these years. His ongoing support, guidance and perseverance have significantly contributed to this final product even after my PhD thesis was approved in April 2013. His useful and insightful suggestions at every stage of writing this book are deeply appreciated.

I am very grateful to Frances Egan and Jack Copeland, the Thesis Examiners. Not only do I thank them for the important comments and suggestions they provided me with, but also for their encouragement to publish an expanded version of the thesis as a monograph.

A number of people have given me feedback on parts of this book or related material in the course of writing my thesis: Joseph Agassi, Cristian Calude, David Chalmers, Eli Dresner, Chris Eliasmith, Luciano Floridi, Gualtiero Piccinini, Giuseppe Primiero, Oron Shagrir, Peter Slezak and Graham White. If I have forgotten anyone, I apologise.

Other people helped me by conversing or corresponding with me on topics related to this book. They include Pieter Adriaans, Philip Cam, Yuri Gurevich, Matthew Johnson, Karl Posch, Matthias Scheutz and Naftali Tishby. Again, I apologise to anyone I have inadvertently omitted.

As a graduate student unknown to them, I wrote or approached Axel Cleeremans, Zenon Pylyshyn, David Rosenthal and Brian Cantwell Smith concerning my research. They generously responded with helpful remarks for which I am grateful. As well, Cantwell Smith's *Age of Significance* project inspired and motivated my research.

I would like to express my gratitude to all the referees (some less anonymous than others) that reviewed my papers for various journals. Their valuable comments, suggestions and criticisms helped me to improve different parts of an earlier version of this book.

Special thanks are also due to Marty J. Wolf for invaluable discussions on the nature of computation and its relation to information. These discussions have encouraged me to explore different avenues and consider other perspectives on the topic at hand. Wolf has also kindly offered many useful comments on a recent version of this book.

Parts of the book, in an earlier form, were presented to various philosophical audiences. I would like to thank those present for their attention, questions and comments.

Thanks are due to the editorial team of the SAPERE series for their assistance during the book's preparation. Lorenzo Magnani made many helpful suggestions on an earlier version of this book as well for his constant encouragement in the process of bringing it to light. Giuseppe Longo provided a detailed report on that version encouraging me to consider some of the main questions from a much broader perspective and to make explicit some implicit assumptions that I have taken for granted. Not but not least, Leontina Di Cecco has handled all the logistics necessary to let the book see the light of day.

Finally, none of this would have been possible without the love and support of my family. I am forever grateful to my wife, Calanit Sabaz, for all her uncompromising and unconditional support during all these years. She has always been by my side, loving and constantly supporting me throughout this challenging journey. Calanit and my daughter, Leah, have always been my greatest inspiration to persevere and work harder, so that I would be comfortable with the results and be able to spend more time with them. Another key motivation for completing this book on time has been the welcome arrival of our second daughter, Noa.

Copyright Statement

Contents

1   **Setting the Stage: Computation in Cognitive Science**   1
     1.1   Introduction   1
     1.2   What is Cognitive Science?   7
     1.3   The Present Thesis in the Landscape of Previous Work   9
     1.4   Formalisms of Computability versus Accounts of Concrete
           Computation   12
     1.5   A Tripartite Division: The Semantic, Causal and Functional
           Views of Computation   15
           1.5.1   Representation, Mental Representations and Computational
                 Representations   15
           1.5.2   The Semantic View of Computation   22
           1.5.3   The Causal View of Computation   23
           1.5.4   The Functional View of Computation   25
     1.6   The Plan of Attack   26
     References   28

2   **An Analysis of the Adequacy Criteria for Evaluating Accounts**
   of Computation   33
     2.1   Introduction   33
     2.2   Smith's Tripartite Adequacy Criteria   34
           2.2.1   The Empirical Criterion   35
           2.2.2   The Conceptual Criterion   35
           2.2.3   The Cognitive Criterion   36
           2.2.4   Asking Too Much and Too Little   37
     2.3   Piccinini's Sexpartite Adequacy Criteria   40
           2.3.1   The Empirical Criterion   41
           2.3.2   The Objectivity Criterion   41
           2.3.3   The Explanation Criterion   41
           2.3.4   The Right Things Compute Criterion   42
           2.3.5   The Miscomputation Criterion   42
           2.3.6   The Taxonomy Criterion   43
           2.3.7   An Adequate Alternative   43

2.4   Von Neumann's Tripartite Adequacy Criteria   48
      2.4.1   The Precision and Reliability Criterion   48
      2.4.2   The Single Error Criterion   49
      2.4.3   The Analogue – Digital Distinction Criterion   49
      2.4.4   A Comparison with Previous Criteria   50
2.5   Implementation Theory Bridging Computability and Concrete
      Computation   51
      2.5.1   The Implementation Adequacy Criterion   51
      2.5.2   The Empirical Criterion and the Cognitive Criterion
              Revisited   52
2.6   Recommended Sexpartite Adequacy Criteria   53
References   55

3   Starting at the Beginning: Turing's Account Examined   57
    3.1   Turing Machines from Mathematical and Physical Perspectives   57
    3.2   The Key Requirements According to Turing's Account   65
    3.3   Turing's Account Evaluated   68
    3.4   Concluding Remarks: Digitality, Determinism and Systems   71
    References   76

4   The Triviality "Account" Examined   79
    4.1   Introduction   79
    4.2   The Searle-Triviality Thesis   80
    4.3   The Putnam-Triviality Theorem   81
    4.4   The Key Requirements for a System to Perform Digital
          Computation   84
    4.5   Trivialisation of Computation Blocked   86
    References   94

5   Semantic Accounts of Computation Examined   97
    5.1   The PSS Account   98
          5.1.1   Introduction   98
          5.1.2   The Key Requirements Implied by the PSS Account   100
          5.1.3   The PSS Account Evaluated   104
    5.2   The FSM Account   107
          5.2.1   Introduction   107
          5.2.2   The Key Requirements Implied by the FSM Account   110
          5.2.3   The FSM Account Evaluated   114
    5.3   A Reconstruction of Smith's Participatory Account   118
          5.3.1   Introduction   118
          5.3.2   The Key Requirements Implied by Smith's Participatory
                  Account   119
          5.3.3   Smith's Participatory Account Evaluated   122

5.4   Arguments for the Semantic View of Computation Criticised........   125
References   131

6   Computation as Information Processing   133
    6.1   Introduction   133
    6.2   Semantic Information, Non-semantic Information and Data   134
          6.2.1   Shannon Information as a Non-semantic Conception
                  of Information   135
          6.2.2   Algorithmic Information as a Non-semantic Conception
                  of Information   136
          6.2.3   Semantic Conceptions of Information.................................   137
    6.3   Features of the Resulting IP Accounts Based on the Different
          Conceptions of Information   141
          6.3.1   A Possible Objection...   141
          6.3.2   Features of the Resulting IP Accounts Based on
                  Non-semantic Information   142
          6.3.3   Features of the Resulting IP Accounts Based on Semantic
                  Information..   145
    6.4   The Key Requirements Implied by the Resulting IP Account   146
    6.5   Problems for the Resulting IP Accounts   153
          6.5.1   Problems for IP Accounts Based on SI or AI.....................   153
          6.5.2   Problems for an IP Account Based on Factual
                  Information..   155
    6.6   The Instructional Information Processing (IIP) Account   157
          6.6.1   An Outline of the IIP Account   158
          6.6.2   The IIP Account Evaluated   160
    6.7   Concluding Remarks..   163
    References   164

7   Causal and Functional Accounts of Computation Examined............   167
    7.1   The Gandy-Sieg Account   167
          7.1.1   Introduction   167
          7.1.2   The Key Requirements Implied by the
                  Gandy-Sieg Account   169
          7.1.3   The Gandy-Sieg Account Evaluated   171
    7.2   The Algorithm Execution Account   175
          7.2.1   Introduction   175
          7.2.2   The Key Requirements Implied by the Algorithm
                  Execution Account   176
          7.2.3   The Algorithm Execution Account Evaluated....................   179
    7.3   The Mechanistic Account..   185
          7.3.1   Introduction   185

7.3.2   The Key Requirements Implied by the Mechanistic
           Account 186
    7.3.3   The Mechanistic Account Evaluated................................. 188
7.4   Concluding Remarks... 193
References 193

8   Computation Revisited in the Context of Cognitive Science **197**
    8.1   A Limited Representational Character of Digital Computation 197
    8.2   Avoiding Ambiguity about Computation..................................... 202
    8.3   The Explanatory Role of Computation in Cognitive Science 207
        8.3.1   Computationalism and Computation Simpliciter 208
            8.3.1.1   Classicism and Symbolic Computation............... 208
            8.3.1.2   Broad Construals of Computationalism and
                       Digital Computation... 209
        8.3.2   Connectionism and Sub-symbolic Computation 211
        8.3.3   Computational Neuroscience and Neural Computation 213
        8.3.4   Extreme Dynamicism and the Non-computational Shift..... 215
        8.3.5   Mechanistic versus Non-mechanistic Explanatory
                   Frameworks.. 217
    8.4   Computation and Cognition – Concluding Remarks...................... 223
    References 226

Chapter 1
Setting the Stage: Computation in Cognitive Science

1.1 Introduction

What is computation? This question may seem, at a first glimpse, trivial and uninteresting. There has already been a long discussion about this subject back in the mid-twentieth century. The discussion produced many extensionally equivalent models of computation, such as Turing machines, lambda calculus, cellular automata, Post machines and recursive functions. The question how such different models of computation *can be* extensionally equivalent is indeed one of the great philosophical questions of our time. However, these are models of abstract computation (or computability, to be more precise), rather than of physical computation. They are, to a large extent, divorced from the details of physical embodiment of computation in real-world systems. Although the question of the equivalence of different models of computability is not addressed in the book, they do play a role in our present enquiry as we shall see below. Our main focus here is on physical – or concrete – computation.

It might be argued that physical realisability of computation is a subject matter for physicists and engineers, but it is of no concern to computer scientists. Theoretical computer science typically deals with functions, mathematical abstractions and algorithms, but not with the physical realisability of those functions and algorithms. Algorithms abstract away the physical reality of the systems implementing them. To enable us to write programs using high-level programming languages, such as Pascal, C, Fortran, C++ and Java, there is a computational hierarchy of abstraction layers at play.

Be that as it may, computer programs, which clearly are of interest to computer scientists, are executed on real-world systems that obey the laws of physics. Besides, programmers are – or at least, should be – also aware of the physical constraints of the physical systems for which their programs are designed. For example, when the CPU of a modern conventional computer consists of multiple cores, multi-threaded programs that are executed on that computer should be

N. Fresco, *Physical Computation and Cognitive Science*,
Studies in Applied Philosophy, Epistemology and Rational Ethics 12,
DOI: 10.1007/978-3-642-41375-9_1, © Springer-Verlag Berlin Heidelberg 2014

designed accordingly to take advantage of this useful computational resource[1]. Also, whereas Turing machines (TMs) are defined over infinite memory, real-world computers always operate with a limited memory (however large it is and even though it can be extended). This suggests that in practice, physical considerations are relevant for computer practitioners too.

Additionally, outside the boundaries of theoretical computer science, computation plays a key role in other disciplines, including psychology, philosophy of mind, linguistics, neuroscience, cognitive science and even physics. Indeed, some physicists and computer scientists claim that Nature computes (cf. Fredkin, 1992; Lloyd, 2006; Wheeler, 1982; Wolfram, 2002; Zuse, 1970). But what sense does it make to wonder whether a table, a waterfall, a strike of lightening, a falling body, an electric current, a growing tree, a quantum dynamic or any natural process computes? (Longo, 2009, p. 62). For these processes to be computational they need to have, at the very least, a starting point (i.e., the input) and an ending point (i.e., the output). The question then remains what computation is physically executed. It seems the view that Nature is a deterministic digital computational process is very problematic (Fresco & Staines, forthcoming). For now, suffice it to say that the claim that Nature computes cannot be even evaluated without understanding what computation means to begin with.

To complicate matters, the claim that cognition – or mind – is computational is commonly made in cognitive science and philosophy of mind. "[Natural] cognition *is* a type of computation" (Pylyshyn, 1984, p. xiii). "It seems to me that we are getting rather close to that situation with respect to the computational theory of mind" (Newell, 1990, p. 56). "[S]ome neuroscientific research on the central system [...] indicates that the mind is computational" (Schneider, 2011, pp. 29–30). "The mind computes answers to problems in a manner analogous to that used by computer software" (Kellogg, 2012, p. 4). But what does it mean to assert that cognition computes? The problem of characterising computation is far from being trivial.

At this point, a brief disclaimer is in order. It is indeed important to "[put] aside, [...] the myths of a computational Universe, of digital calculating brains, of genetic 'programs'" (Longo, 2009, p. 69). However, the present enquiry focuses primarily on understanding what it means precisely to claim that cognition computes. This enquiry shall proceed for the most part without either a commitment to or a denial of the computationalist thesis[2]. The computationalist

---

[1] In multi-core processors (such as Intel's quad-core series i3, i5, and i7 and AMD Phenom II X4) the CPU has two or more independent processors (each called a *core*). The performance improvement gained by the use of this technology depends on the programs using it. This is particularly true in multi-threaded programs, in which a program consists of multiple threads running simultaneously and sharing the same memory address space. A thread is a flow of control (consisting of a sequence of instructions) within a process that is scheduled for execution by the operating system on the computer processor.

[2] Computationalism is roughly the view that cognition is a type of computation. Computationalism and all its varieties are examined in detail in Chapter 8.

thesis crucially depends on the particular interpretation adopted of computation. Only once it is clear what the computationalist takes a computing system to be, can her thesis be judged as being either true or false.

Some researchers are quick to employ the notion of computation *simpliciter* when asserting basically that cognitive activities are computational. Computation simpliciter refers to generic computation that encompasses digital computation, analogue computation, neural computation, quantum computation and any other form of "computation" (Piccinini & Scarantino, 2011, pp. 10–13). Unfortunately, it seems that a clearer understanding of computation is often distorted by philosophical concerns about cognition. Many researchers in referring to *digital* computation simply adhere to the classical notion of *computability* when attempting to explain cognitive behaviour computationally. Classical computability theory originated with the seminal work of Kurt Gödel, Alonzo Church, Alan Turing, Stephen Kleene and Emil Post in the 1930's. It studies what functions on the natural numbers are computable, and not the spatiotemporal constraints that are inherent to natural cognitive phenomena.

Any analysis of cognitive phenomena that is based solely on mathematical formalisms of computability is at best incomplete. It has been proven that Post machines, Kleene's formal systems model, Gödel's recursive functions model, Church's lambda calculus, and TMs – are extensionally equivalent. They all identify the same class of functions, in terms of the sets of arguments and values that they determine, as computable (Kleene, 2002, pp. 232–233). Since what is computed is expressed by these formalisms in terms of functions from inputs to outputs, these formalisms can be compared according to the class of functions they compute (Scheutz, 1999, p. 164).

However, concrete digital computation as it is actualised in physical systems seems to be a more appropriate candidate for the job of explaining cognitive phenomena. It is not in vain that some of the reigning trends in contemporary cognitive science, be that connectionism[3] or dynamicism[4], emphasise the embeddedness and embodiment of cognitive agents. This is one motivation for examining extant accounts of computation that are not limited to an analysis of computability, before we can make any sense of talk about 'cognitive computation', 'neural computation' or 'biological computation'. Since our interest is in physical computing systems, human-engineered computing systems such as conventional digital computers and logic circuits are used throughout this book to

---

[3] Connectionism is one of the main paradigms in computational cognitive science that views cognition as sub-symbolic computation. Connectionists argue that cognition should be explained by neural network activity and thus should be modelled using artificial neural networks, also known as connectionist networks. Connectionism is discussed in more detail in Chapter 8.

[4] Dynamicism is a relatively recent paradigm in cognitive science that views cognition as a complex, ongoing dynamical interaction of the cogniser with its environment. Dynamicism often presents itself as a departure from orthodox computational cognitive science. It is discussed further in Chapter 8.

exemplify various points. These examples do not limit the generality of the arguments put forward and are used as paradigmatic cases of concrete computation.

For the sake of clarity, some distinction among different types of computation may be useful here. A good starting point is Bruce MacLennan's distinction among *offline* computation, *embedded* computation and *embodied* computation (2011, p. 228). But this distinction lacks an important type, which stands in direct contrast with offline computation, namely, *online* computation[5]. Embodied computation exceeds the scope of this book and is only addressed indirectly. It remains an open question though whether embodied computation is indeed the type of computation that takes place in nature. But at least traditionally, it has not been invoked as the basis of the Computational Theory of Mind (CTM).[6]

In offline computation any interaction with the environment is best characterised as input-process-output. The physical input, such as punched cards or magnetic tape, is presented to the computing system and converted into some specific computational representation and the output generated is converted back into specific physical representation (e.g., punched cards, magnetic tape or printed-paper). Three characteristics of offline computation are immediately clear. First, the output generated can be used at a later stage once the program execution has terminated, provided that one deals with *program executing systems*. Second, the conversion of physical input into some computational representation is an abstraction procedure that enables the computation process. Third, following from the previous characteristic, offline computation can be construed as the evaluation of a mathematical function on some given argument(s) as in the classical theory of computability (ibid).

An online, or interactive, computational process, in contrast, interacts with the environment after the initial input was entered. One possible informal characterisation of an ordinary interactive algorithm is that it cannot successfully complete a step while there is an unanswered query from that step[7]. The only data that the algorithm receives from the environment during a particular step consist of the replies to the queries issued during the step (Gurevich, 2005, p. 35). Examples of online computation are digital computers that communicate with external databases or other servers for the execution of some tasks and computer game models, each of which can be thought of as a series of moves between a player and an opponent, or the system and the environment.

Online computation differs from its offline counterpart, in which the process begins with some input data and proceeds without any further interaction with any external system (Soare, 2009, p. 387). In online computation there is typically a

---

[5] As its name suggests, online computation is performed in an online (i.e., connected) mode. The computing system remains connected to some input device, as the input is entered in a piecemeal fashion.

[6] 'CTM' and 'computationalism' are used interchangeably throughout the book.

[7] A *general* interactive algorithm, on the other hand, may finish a step even in the absence of some replies to the queries issued by the algorithm (Gurevich, 2005, p. 36).

fixed algorithm or procedure at its core, but there is also a mechanism for the process to communicate with its environment[8]. Still, as in its offline counterpart, despite the interaction with the environment, the online computational process is executed on some computational medium.

Embedded computation is a specific subclass of online computation. This type of computation is typically non-terminating. Its physical size and real-time speed have to be compatible with the physical system in which the computing device is embedded. Further, the basic structure of embedded computing systems consists of sensors, controller(s) and actuators. These systems are highly interactive with their environments through real-time feedback loops from the actuators back to the sensors. Still, the sensors and actuators perform conversions to and from the computational medium typically independently of any specific physical realisation (MacLennan, 2011, p. 228).

Unlike the other types of computation mentioned above, embodied computation involves either little or no abstract computation. Embodied computation should be understood as a physical process in an ongoing interaction with its environment. Computational processes emerge as regularities in the dynamics of a physical system that allow the system to fulfil its function. In embodied computation, feedback may be implemented by naturally occurring physical processes, such as evaporation and degradation of chemicals or diffusion in fluids. In such processes, which occur anyway, the computation is performed "for free" by the implementing physical substrate. As well, in natural physical systems randomness comes for "free" in the form of noise, uncertainty, imprecision and other stochastic phenomena (MacLennan, 2011, pp. 228–230).

In the light of the above considerations, it seems that the time has come for the phenomenon of concrete digital computation to be attended to. As well, cognitive scientists should be explicit about what they take 'computation' to be when employing it in the context of cognitive explanations. Of course, this need is even more pressing, if cognitive computation is indeed a type of, or overlaps with, digital computation (or, perhaps, a type of embodied computation). If it is not, then it is important to understand why not and in what respects it is different. This has consequences not only for the resulting thesis underlying CTM, but also for the tools available to analyse this type of computation. The best way to get some insight into this matter is to examine possible answers to the problem of what concrete digital computation is.

Since the focus of the present work is on physical computation, we should make a few brief observations about the physical theory underlying the present analysis. Modern physics is typically divided into two parts: Classical Physics and Quantum Physics. They do not seem to be unified. The former deals with macroscopic bodies, whereas the latter deals with microscopic objects. The

---

[8] Incidentally, Robert Soare argues that whilst the fixed algorithm may be identified with some conventional TM where the communication mechanism is coded into integers, it is better regarded as an oracle TM (2009, p. 387). This point and the notion of an oracle TM are briefly discussed in Chapter 3.

question to be asked then is whether we have to pay attention to classical or quantum physics. In the subatomic world, which includes atoms, atomic nuclei and even smaller fundamental particles, one should pay attention to quantum physics. But in the large-scale world classical physics is more adequate for analysing the phenomenon in question. As a simple example, let us consider the analysis of a glassful of water. Water is composed of H_2O molecules, yet this molecular structure is irrelevant to a physicist analysing the flow of water, its turbulence, its pressure, etc. On the other hand, when the same physicist wishes to analyse a single H_2O molecule, which is a quantum entity, she enters into the realm of quantum physics (Ford, 2011, pp. 5–6).

Similarly, since our present analysis concerns macroscopic computing systems, it shall be conducted mostly in the context of classical physics. This is not to say that quantum physics is any less valid in our large-scale world than in the subatomic world. Many properties of a large-scale system, such as how well it conducts electricity or how much heat is needed to raise the system's temperature 1 degree, can be understood only in terms of quantum properties. However, the quantum underpinnings of such large-scale behaviour are typically hidden from direct observation (Ford, 2011, p. 6). There is no conflict between the two ways of doing physics, for quantum physics, whether evident or not, is always at work in every domain. Yet, it is classical physics that deals with force and motion, heat and entropy (topics that are mentioned again in Chapters 3, 4 and 6) and electricity. Classical physics is successful in the realms in which it applies (Ford, 2011, p. 7). Using classical physics, the hardware of conventional computing systems can be analysed, for example, at the macroscopic level at which registers change their stored data or at the level at which logic gates operate on various input lines. Using quantum physics, the same hardware can certainly be analysed further at the subatomic level, but then it becomes increasingly harder to trace the high-level operations that the computing system performs. Quantum computation is a fascinating topic in its own right, but it exceeds the scope of this book. Unless specified otherwise, the reader should assume that the underlying physics used here is classical.

The remainder of this chapter is organised as follows. In Section 1.2, we present the context in which concrete digital computation is analysed, namely cognitive science. This multidisciplinary scientific research can be construed broadly or narrowly, but this book deals mostly with the latter construal. Subsequently, in Section 1.3, we consider a sample of work that has dealt with this problem either directly or indirectly. In Section 1.4, we argue that formalisms of computability may indeed provide the mathematical tools required for determining the plausibility of *computational level theories*. But if cognition could be explained computationally, it would be *concrete* computation rather than computability at the basis of the explanans. This chapter continues with a critical survey of the notions of representation that are employed in cognitive science and those that may play a role in concrete digital computation proper (i.e., not cognitive computation or biological computation). It then outlines the tripartite division amongst the main views of computation. The concluding section outlines the structure of this book.

As will become apparent in due course, the objective of this book is twofold. First, it is to promote a clearer understanding of concrete computation in the cognitive science discourse whether one embraces CTM or rejects it. Second, it is to establish the foundational role of concrete computation in cognitive science whilst rejecting the extrinsically representational[9] character of computation proper.

1.2 What is Cognitive Science?

One would think that there is a simple answer to the question what cognitive science is. It is the scientific investigation of cognition. The answer, however, is not that simple. For one thing, in the early days of cognitive science, personality, motivation and emotion, which seem plausibly cognitive, were regarded as key topics in their own right (Boden, 2008, p. 1946). More crucially, cognitive science can be construed narrowly or broadly and this has implications not only for the disciplines partaking in this science, but also for the subject matter itself. When construed narrowly, it typically means computational cognitive science. When construed broadly, it encompasses many disciplines and research programs that do not necessarily commit to cognition being computational.

For the first few decades, cognitive science was identified with CTM. Ned Block, for instance, identified cognitive science with the "cognitivist ideology", whose metaphor is that of cognition as a digital computer. "When I speak of cognitive science, I have in mind an *ideology*... presupposed by a significant line of work in cognitive psychology and artificial intelligence, and to some extent linguistics and philosophy [...] Many writers on this topic prefer to use 'cognitivism' to refer to [this] ideology" (Block, 1983, p. 521, fn 17, italics added). Similarly, John Searle claims that "the basic assumption behind cognitive science is that the brain is a computer and many processes are computational" (1992, p. 198). A similar line is taken by Steven Pinker in stating that "the key idea of cognitive revolution of the 1950s and 1960s [was] that the mind is a computational system" (2005, p. 1).

This assumption clearly was and still is fundamental in the work of many cognitive scientists. George Miller, a psychologist and one of the founders of the field, describes the original dream of cognitive science as "a unified science that would discover the representational and computational capacities of the human mind and their structural and functional realisation in the human brain" (2003, p. 144). Unsurprisingly, in a progress report by computer scientists Marvin Minsky and Seymour Papert from 1972 they summarised their view as follows. "Thinking is based on the use of symbolic descriptions and description-manipulating processes to represent a variety of kinds of knowledge [...] The ability to solve new

---

[9] 'Extrinsic representations' are explicated in Section 1.5.1.

problems ultimately requires the intelligent agent to conceive, debug and execute new [computational] procedures" (Minsky & Papert, 1972, p. 2). Later in the nineties this remained a working assumption in cognitive science (Pylyshyn, 1999, p. 10). And it is certainly proving to be very useful in computational modelling in neuroscience (Cleeremans, 2005, p. 81; Decety & Lamm, 2007, p. 580).

However, when cognitive science is narrowly construed as *computational* cognitive science, showing that cognition is not computation would shake the very foundation of the overall research program. It is generally agreed that cognitive science was founded because researchers from different disciplines realised that they were all interested in studying intelligence or cognition using similar theoretical constructs, namely, the notions of representation and computation. Admittedly, they also realised that to accomplish this common goal a multidisciplinary effort was required (von Eckardt, 2001, pp. 457–458). But presuppositions about representation and computation have been the source of much of the criticism unleashed against CTM for several decades.

Cognitive science is no longer limited to the traditional aspects of cognition that seem to share the key feature of being underpinned by mental representations. Traditionally, the focus of cognitive science was on problem solving, learning, reasoning, attention, memory, perception and language processing. Percepts are typically about the objects that cause them. Real memories are based on certain perceptions. Reasoning is also about ideas representing the way the actual and possible world is and could be, respectively (Harnish, 2002, p. 5). But at later stages, other topics also became the focus of some specific disciplines in cognitive science, including emotional intelligence, social interaction, sensorimotor control, interaction with the environment, orientation and adaptation.

Consequently, contemporary cognitive science is a multi-disciplinary research area that would not simply fail, if cognition turned out to be non-computational. Cognitive psychology examines the typical human, individual differences and group differences. Clinical psychology concentrates on psychopathology. Biology and psychology analyse animal cognition (von Eckardt, 2001, p. 459). Anthropology studies cognitive phenomena from a cross-cultural perspective. Cognitive neuroscience tells us how systems with specific neural features give rise to perception, action, memory, language and other cognitive functions. Linguistics is dedicated to language learnability, language change and the competence (i.e., the speaker-hearer's knowledge of her language) versus performance (i.e., the actual usage of language) distinction (Chomsky, 1969, p. 4). Computer science provides the most complete understanding of how a complicated capacity can be realised in terms of software and hardware. Artificial intelligence studies machine intelligence and models of human cognition with the overall goal of building autonomous intelligent agents.[10] The dominance of the last two disciplines would certainly diminish with a realisation that cognition is *not* computation.

---

[10] Yet, some researchers simply study machine learning from a purely engineering perspective with a more modest goal of producing better computer programs, rather than studying human cognition.

This book is not strictly limited to the narrow construal of cognitive science. However, since its main objective is promoting a clearer understanding of concrete computation, it has consequences primarily for computational cognitive science. Whilst our focus is mainly on computationalism, narrowly and broadly construed, as well as connectionism, the morals to be drawn here apply to any cognitive scientist making assumptions about or claims based on computation.

1.3 The Present Thesis in the Landscape of Previous Work

The literature contains many attempts to clarify the notions of computation simpliciter and digital computation, in particular. Matthias Scheutz, for example, argues that there is no satisfactory account of implementation to answer questions critical for computational cognitive science (1999, p. 162). He does not offer a new account of concrete computation *per se*. Instead, he suggests approaching the implementational issue by starting with physical digital systems and progressively abstracting away from some of their physical properties until a mathematical description remains of the function realised.

In a similar vein, David Chalmers (2011) focuses on the implementational issue, only to offer another formalism of *computability*. His proposed formalism is based on combinatorial state automata (CSA) supplanting the traditional finite state automata (FSA). Chalmers also argues that a theory of implementation is crucial for digital computation to establish its foundational role in cognitive science. The main motivation behind both Scheutz and Chalmers' efforts to clarify the notion of implementation is to block attempts by Hilary Putnam, Searle and anyone else to trivialise computation.

Other notable discussions of computation in cognitive science include David Israel (2002), Oron Shagrir (2006, 2012), Jack Copeland (1996), Robert Matthews (2011), Kenneth Aizawa (2010), MacLennan (1999, 2004, 2011), Gualtiero Piccinini (Piccinini & Scarantino, 2011; 2006, 2007), Brian Cantwell Smith (2002, 2010) and the list continues. Israel claims that often it seems that a better understanding of computation is hampered by philosophical concerns about mind or cognition (2002). Yet, "[o]ne would, alas, have been surprised at how quick and superficial such a regard [to computation] has been" (Israel, 2002, p. 181). Shagrir examines a variety of individuating conditions of computation showing that most of them are inadequate, for they are either too narrow or too wide (2006). Although he does not provide a definitive answer as to what concrete computation is, Shagrir points out that connectionism, neural computation and computational neuroscience are incompatible with the widespread assumption that digital computation is executed over representations with combinatorial structure. Copeland defends Turing's analysis of computation and seeks to uphold its sufficiency (1996). In doing so, he provides a detailed analysis of what it means for some system to *compute a function*. Copeland's algorithm execution account is examined in Chapter 7.

Matthews adopts a different approach in characterising *natural* computation, rather than computation by *artificial* systems (2011). On his view, most computer scientists would agree that what gets computed in the course of a computation is the value of some mathematical function. Matthews argues that there is no reason to assume that this function is computed in the brain in the same way that it is computed by conventional computers, namely by means of following some algorithm. Nevertheless, computational explanations in cognitive science have a mathematical function-theoretic character that need not rely on algorithms.

Matthews does not settle the question regarding concrete computation broadly construed. Instead, he draws our attention to the "perspectival property that [artificial] computers have in virtue of the way we users construe them" (Matthews, 2011). On his view, a key criterion for an artificial system to be deemed computational is the utility of the computed value for the *user*, but for *natural* computing systems it is the utility of the computed value for the *system*. However, it is not at all clear why one would insist on classifying cognitive activities as computational when all they share with computing systems proper is the specification of a mathematical function being computed.

Aizawa argues that we should not seek a single set of necessary and sufficient conditions for individuating computation that would fit all uses including those of computer science and cognitive science (2010). But he also argues that, fundamentally, we should reject the ongoing focus on *Turing-equivalent* computation. Aizawa emphasises that in philosophy of mind, computation is often understood as mere symbol manipulation, rather than as a Turing-like conception. Particularly in the context of CTM debates regarding the systematicity of mental representations, the wrong inference is made that a Classical architecture is a Turing-equivalent formalism.

On the other hand, in certain areas of neuroscience there is another notion of computation in play (Aizawa, 2010). For instance, in some research on functional maps of the brain, the basic idea is that some, but not all, maps are computed in the sense of an arithmetic function of its inputs in a way that is quite removed from the Turing-like conception. However, as it is argued below, TMs and other equivalent formalisms of computability are of the *wrong* kind to individuate concrete computing systems – natural or not. For the same reason, it is wrong to equate any Classical architecture with a Turing-equivalent formalism to begin with. A standard TM is a synchronous computing system, whereas, nowadays, asynchronous machine computation is widespread. Furthermore, the TM model deliberately ignores spatiotemporal considerations, whilst the Classical architecture does not. Lastly, the TM is a good model for analysing computational complexity, but it is not very satisfactory for analysing modularity. (In passing, it should be mentioned that the lambda calculus fares better in that regard.)

Importantly, three notable, and relatively uncommon, examples of genuine attempts to explicate the notion of computation are MacLennan's, Piccinini's and Smith's. MacLennan's analysis of computation is primarily concerned with its manifestation in natural processes. Whilst he focuses on the physical aspects of computation, his main interest is in *embodied* computation. MacLennan argues that the Church-Turing model of computation is not well suited to address issues in computation occurring in nature (2004, p. 116, 2011, p. 250). By his lights, the end of Moore's law is in sight. The physical limits to the density of binary logic devices and to their speed of operation will require us to approach computation in new ways. Increasing the speed and density of computation will force it to take place on a spatiotemporal scale nearing that of the underlying physical substrate. As a result, post-Moore's-law computation will increasingly depend on a considerable assimilation of computation to physics.

Accordingly, MacLennan argues that we should seek a deeper and broader concept of computation in both natural and artificial systems (2011, pp. 225–226). Rather than relying on the narrow conception of TM computation, he suggests adopting *continuous* computation, which is more compatible with processes occurring in natural systems (MacLennan, 1999, 2004, pp. 127–130). However, the character of the computation type that MacLennan endorses implies that it is *analogue* rather than *digital* computation. Our present focus here is on digital computation, which is prominently employed by computationalists and others in cognitive science (cf. Fodor & Pylyshyn, 1988; Haugeland, 1985; Newell & Simon, 1976; Pylyshyn, 1989; van Rooij, 2008; Von Eckardt, 1993).

Furthermore, Piccinini demonstrates how on various readings of computation, some have argued that computational explanation is applicable to psychology, but not, for instance, to neuroscience (2006). Still, many neuroscientists routinely appeal to computations in their research. Elsewhere, Piccinini examines the implications of different types of digital computation, as well as their extensions' relations of class inclusion, for CTM (Piccinini & Scarantino, 2011). He proposes a mechanistic account of digital computation that aims to overcome the weaknesses of other accounts in the literature (Piccinini, 2007). The mechanistic account is examined in detail in Chapter 7.

But it seems that nobody else in the literature has ever undertaken a more ambitious project than Smith to systematically examine the extant accounts of computation and their role in both computer science and cognitive science. In his "The foundations of computing", Smith lists the following six construals of computation: Formal Symbol Manipulation, Effective Computability, Algorithm Execution, Digital State Machines, Information Processing and Physical Symbol Systems (2002). His *Age of Significance* project, which is now long coming, aims to shed some light on the murky notion of computation, putting each one of these construals under careful scrutiny (Smith, 2010).

The main underlying thesis of this book is weaker than Smith's negative conclusion. Surprisingly enough, Smith concludes that there is *no* adequate account of computation and there *never* will be one. For computers *per se* are not "a genuine subject matter: they are not sufficiently special" (Smith, 2010, p. 38). *Pace* Smith, it is argued here that there is no compelling reason to reject *all* accounts of computation as inadequate, let alone to preclude the possibility of ever coming up with an adequate account. Indeed, not only are extant accounts of concrete digital computation non-equivalent, but most of them are *inadequate*. In due course, we shall see that these accounts are not only intensionally different – this is quite trivially the case, but also extensionally different. This already suggests that at least some of them are inadequate for singling out the wrong class of digital computing systems.

1.4 Formalisms of Computability versus Accounts of Concrete Computation

There is no question whether mathematical formalisms of computability are adequate analyses of *computability*, but they are of the wrong kind to individuate concrete computing systems. Formalisms of computability certainly provide the mathematical tools required for determining the plausibility of *computational level theories*.[11] Nevertheless, any particular formalism does not specify the relationship between *abstract* and *concrete* computation. It is at the physical level that the algorithm is specified and bound by the implementing physical system. And it is the physical architecture of the computing system that determines the very nature of the program implementing that algorithm. (That is, provided that the system in question executes a program, since it is not clear that a conventional OR gate, for example, executes a program.)

This argument follows almost immediately from David Marr's proposed three levels of analysis. If his explanatory model is interpreted as a top down methodology[12], then the top level provides an extensional account of *what* is computed, the middle level specifies *how* it is computed, and the bottom level describes the actual implementation. Marr's top/computational level characterises the mathematical (or cognitive, in the context of cognitive systems) input/output function being computed. The middle/algorithmic level specifies how the system computes this function and is consistent with various formalisms of computability. The implementational/bottom level shows how the system works physically. It is at this level that an account of concrete computation is needed to supplement the analysis enabled by the formalism of computability.

Indeed, formalisms of computability provide the mathematical tools required for evaluating the plausibility of computational level theories of cognition. Since our cognitive capacities are constrained by the fact that we are finite systems with

---

[11] A computational level theory is, typically, intended to characterise the *what* and *how* a given system computes. The *what* element characterises the function that the system computes and the *how* element specifies the algorithm used to compute this function (Shagrir, 2010, p. 477).

[12] This conventional interpretation has had its fair share of criticisms (see, for example, Bell et al., 2001, pp. 209–212; Dennett, 1998, pp. 249–260; Shagrir, 2010). For our purposes, suffice it to say that Marr identifies the top level as *computational*. But it is his algorithmic level that is on a par with formalisms of *computability*, and the implementational level that describes *physical computation*.

limited resources for computation, some argue that classical computability theory and complexity theory can assist cognitive science by providing theoretical constraints on the set of feasible computational level theories. Moreover, it may be shown using a particular formalism that some cognitive function is intractable. An underlying assumption of this view is that a cognitive capacity involves the effective computation of a specific input/output function F (i.e., a set of ordered pairs), which given an initial/input state $i \in I$, realises a final/output state $o = F(i)$ ($o \in O$) as specified by function $F: I{\rightarrow}O$. If a computational level theory of some cognitive capacity is assumed veridical, then further attempts can be made by scientists to analyse it at the algorithmic and the implementational levels (van Rooij, 2008, pp. 939–941).

Once a particular mathematical formalism was chosen to evaluate the feasibility of a computational level theory, the next level of abstraction has to be taken into account, namely the physical level. Each one of the mathematical formalisms of computability provides a definition of the set of allowable operations used to compute functions and their associated computational costs by implication. So, in order to determine the computational resources that are required, the particular formalism needs to be specified, though some complexity distinctions, such as between P and NP running time, are insensitive to the particular formalism of computability. An algorithm is said to be P, or *polynomial* time, if its execution time is upper bounded by a polynomial in the size of the input for the algorithm, i.e., $T(n) = O(n^k)$ for a constant $k \in \mathbb{N}^+$. Formally, a language L is in the class NP (*non-deterministic polynomial* time) iff L is accepted by a non-deterministic TM, which operates in polynomial time (Karp, 1972, p. 91). Nonetheless, the particular formalism does not specify the relationship between *abstract* and *concrete* computation.

If we are to agree with Marr that problems must first be faced at the computational level, then we cannot avoid questions regarding *algorithmic* and *concrete computation* at the next levels. Daniel Dennett suggests that an analysis at the computational level "might be nothing more that the demand for enough rigour and precision to set the problem for the next level down, Marr's algorithmic level" (1998, p. 232). *Some* formalisms of computability, such as the TM and the lambda calculus, specify the *algorithm* to be employed to compute the computational level function, whereas others, such as recursive functions, do not. The *appropriate* formalism provides an algorithmic level explanation of *how* ordered pairings at the computational level are computed. However, there could be many *algorithms,* which correspond to any particular computational level function, that could, in turn, be implemented in any number of different ways.

So, it is only at the implementational level that the algorithm has been specified and constrained by a particular physical substrate. This is where an account of concrete computation has to be specified explicitly. Any enquiry concerning physical computing systems should include what we know about the system's architecture and how it shapes the algorithms the system is running. The appropriate account of concrete computation in conjunction with the relevant formalism of computability can explain both the *what-* and *how-* questions about the computation carried out by the system concerned. This also shows that there is no conflict in subscribing simultaneously to both a formalism of computability and an account of concrete computation.

An analysis in terms of, say, TMs does not account for spatiotemporal considerations as already remarked above. The TM model and likewise other formalisms of computability were developed to address issues in effective computability and formalist approaches to mathematics. Within this context it certainly makes sense to deem something computable, if it can be computed in a finite number of steps (but of indeterminate time duration) using a finite (but, in principle, unbounded) amount of memory. Computability is defined in terms of sets of functions. However, formalisms of computability ignore real-time rates of operations, robustness in the presence of noise and errors, physical size, consumption of energy and so on (MacLennan, 2011, p. 250).

Accounts of *concrete* computation, on the other hand, should not only account for the physical level of computing systems, but also make provisions for their algorithmic level. For example, the algorithm execution account in the version proposed by Copeland (see Chapter 7) builds on Turing's analysis at the algorithmic level. Yet, it extends Turing's analysis to be closely coupled with the physical level in considering the formal description of the computing system, the algorithm executed and the supporting architecture. Another example is Smith's participatory account (see Chapter 5) that emphasises both the algorithmic and physical levels in explaining the crossing of boundaries between abstract algorithms and real-world computing systems.

Furthermore, Robin Gandy argued that concrete computation requires an independent analysis that considers the limits of physical computation, unlike Turing's analysis (1980, pp. 124–125). Gandy postulated two explicit physical constraints on computing systems. The first one was a lower bound on the size of distinguishable atomic components of the computing system. The second constraint was an upper bound on velocity of propagation of information (implicitly postulating a key role for *information* in digital computation). These two constraints were motivated by purely *physical* considerations and are discussed further under the Gandy-Sieg account of computation in Chapter 7.

Lastly, when digital computation is only defined abstractly, say, in terms of TMs or Post machines, difficulties arise in developing a satisfactory account of implementation (Scheutz, 1999, p. 170). A more practical approach is to define computation in terms of an abstraction over the physical properties determining the functionality of the physical computing system. Computer practitioners take this approach when they define programming languages for the hardware constructed in the computer industry to make it usable for its users. By abstracting over specific hardware particulars, such as the speed of gates and CPU clocking, they can provide an abstract description of what it takes to compute on some physical machine. In this way computation and implementation are defined together.

1.5 A Tripartite Division: The Semantic, Causal and Functional Views of Computation

The accounts of concrete digital computation under analysis in the book can be crudely classified into one of the following three main views of computation.

I.   The semantic view of computation, according to which computation is individuated by its semantic properties. On this view, computational individuation makes an essential reference to representation and content.

II. The causal view of computation, according to which computation is individuated by its causal properties. On this view, the formal structure of computation is mirrored by the causal structure of its implementation without appealing to extrinsically representational properties.

III. The functional view of computation, according to which computation is individuated by its functional/organisational properties (as a special case of causal properties). On this view, functional/organisational properties are insensitive to content and need not presuppose extrinsically representational content.

Before turning to examine these three types of view, some clarification is required if we are to understand the contentious claim that computation proper is essentially representational, in the extrinsic sense. This claim is an underlying assumption at the heart of CTM. Its immediate implication is that theories of computation and theories of mind and cognition become closely coupled. To make sense of this claim we should first lay out some clarifying observations about representation and mental representation.

1.5.1 Representation, Mental Representations and Computational Representations

As was pointed out by William Ramsey, cognitive states and structures are often characterised as representational without a comprehensive analysis of 'representation' (2007, p. 8). Cognitive scientists sometimes mistakenly assume that there is some pre-theoretic consensus on what representation actually involves. Most attempts to provide all-encompassing definitions of representation that are then used as criteria for determining whether or not a theory invoking representations is justified in doing so have failed. For most definitions give rise to a number of counter-examples showing that the definitions are either too permissive or too strict. Instead, Ramsey proposes to treat 'representation' as a disjunctive concept in accord with Ludwig Wittgenstein's family-resemblance idea. Whether a state or structure *is* representational becomes a case by case judgment call by exploring how a representational posit is said to operate in a particular system and evaluating its role in terms of our intuitive conception of representation (Ramsey, 2007, pp. 9–11).

In an attempt to gain a better understanding of mental representation, Barbara von Eckardt first examined the nature of *non-mental* representations developing Charles S. Peirce's analysis. She adopted Peirce's triadic relation among the represented object, the representing vehicle (Peirce's *representamen*) and the interpretation (Peirce's interpretant[13]) (Von Eckardt, 1993, p. 145).[14] The represented object could be anything such as a physical object, a relation, a state of affairs or a property. The representing vehicle could be, for example, a map, a painting, a photo, a script or a spoken word. Both the represented object and the representing vehicle are, at least in principle, objective and publicly verifiable, whereas the interpretant introduces a subjective knower-dependent element. On this view, R is a non-mental representation only if R represents an object with respect to an interpretant (Von Eckardt, 1993, p. 148).

Still, it is *mental* representation that plays a key role in philosophy of mind and cognitive science; so it is important to, at least briefly, examine it to understand the motivation behind insisting on representation being *essential* to computation proper. When cognition is viewed as being involved in coordinating the behaviour of a cognitive agent in its environment, one plausible strategy is to view some of its internal states and processes as carrying information about or *standing in* for those relevant aspects of its body and external states of affairs in negotiating its environment (Bechtel, 1998, p. 297). Mental representations function as Stand-Ins for objects or events outside the cognitive agent and once the agent obtains those representations, it can operate on them rather than needing the actual objects or events (Fodor, 1980).

Mental representations have two important features, namely being *physically realisable* and being *intentional*. The first feature implies that they have causal powers. And being intentional, mental representations convey meaning or content. This characterisation presupposes a distinction between the *vehicle* (a physical state or structure, such as a string of symbols) and its *contents*. The issue concerning the admissible *vehicles* of representation remains highly controversial (Egan, 2011). Yet, it is common in computational cognitive science to assume that these vehicles are computational structures or states in the brain (Von Eckardt, 1993, pp. 168–169). This traditional notion of representation allows for the possibility of a *misrepresentation*.

---

[13] An interpretant of a non-mental representation for Peirce is the mental effect on an interpreter (typically, the interpreter's thinking about the represented object) for whom the representation holds (Von Eckardt, 1993, p. 148). Whether this interpreter is actual or merely possible is yet another matter.

[14] This triadic scheme of representation is well entrenched in philosophy of mind. Peter Slezak, for one, criticises this scheme for inviting an assimilation of internal and external representations and encouraging the postulation of the notorious homunculus (2002, p. 240). Yet, any attempt to reduce the triadic relation to a dyadic one yields intractable problems.

To misrepresent is to exercise the capacity to represent something as being so when it is not in fact so. Only if a system has the capacity to misrepresent, that is, to get things wrong, does it have the power to get things right (Dretske, 1988, pp. 65–67). The power of indicators to represent is relative to what they are *intended* to indicate regardless of their success in fulfilling this function.[15] They are intended to indicate something, because we assign this function of indication relative to some standards. It is humans (i.e., systems that are the *source* of intentionality, on this view) who assign the functions that determine the representational capacity of the system. What makes it possible for a map to misrepresent the geography of a particular area is the symbol token (e.g., a blue mark on the paper) failing to carry the information it is its job to carry (e.g., there exists a lake in the park along these coordinates). This token does this job in virtue of its type whose information-carrying role is assigned by the producer(s) of the map and accepted by its consumer(s) (Dretske, 1981, pp. 192–193).

Furthermore, according to the traditional view of mental representations, some naturalistic relation must obtain between the representee and the representer (Egan, 2011). This naturalistic relation was also what Peirce had in mind in introducing his triadic relation (Von Eckardt, 1993, p. 159). However, at least at a first glance, it seems that this naturalistic relation principle is violated, if the interpretant is taken to be the thought of some knower. One plausible solution is that this semantic content-determining relation, which holds between the mental structure and the represented object, has to be specified, or be grounded, in non-intentional and non-semantic terms (Von Eckardt, 1993, p. 189). According to Frances Egan, the motivation for insisting that this relation be naturalistic is to allow computational cognitive science to provide a fully mechanical account of cognition and provide a solid basis for a naturalistic account of cognitive capacities (2011).

More specifically, there are two main approaches in computational cognitive science to the interpretation of representational vehicles. According to computationalism (more precisely, classicism, as we shall see in Chapter 8), complex data structures, formally construed, constitute the representational vehicles of our mental representations. According to connectionism, the representational vehicles are either local, that is, the individual activated connectionist neurones, or distributed, that is, sets of activated connectionist neurones (Von Eckardt, 1993, pp. 169, 176). Whilst the main motivation for taking the computationalist's data structures to be the bearers of representational content is their compositionality-enabling structure, connectionist (or neural) networks do not straightforwardly exhibit such structure.

Connectionist networks consist of multiple interconnected homogenous units called 'neurones'. These networks can be classified into two general categories:

---

[15] To be sure, representation and indication are similar but not equivalent concepts. For a discussion of their similarities and differences, see Cummins and Poirier (2004).

feedforward networks and feedback, or recurrent, networks. In the former case, neurones are arranged across multiple layers such that the activation of neurones in layer n depends only on the outputs of neurones in layer number n-1. Neurones that are neither input nor output neurones, but intermediate, are called hidden neurones. Hidden neurones in layer n have edges from neurones in layer n-1 and to neurones in layer n+1. Layers consisting of hidden neurones are called hidden layers, though, strictly, a feedforward network need not have any hidden layers. The outputs of neurones are updated starting with the *input layer* and then layerwise to the last one being the *output layer*. In the case of recurrent networks, the output of a given neurone at time t_i can causally influence its activation state at t_j via feedback loops in the network. In those networks there is no forward direction in the flow of processing, but rather a "relaxation" in which the connectionist neurones are updated until a fixed point is reached.[16]

There is some controversy regarding the nature of representations in connectionist networks. On the localist interpretation, each individual neurone, which is active in a particular distributed activation pattern, realises an individual representation contributing to the overall content of the activation pattern. On the distributive interpretation, a representation is realised by either an activation pattern or an activation pattern *and* its weighted connections.

Moreover, an important distinction to be drawn regarding representations in this context is between processes operating *on* representations and representations *figuring in* processes (Bechtel, 1998, pp. 299–300). The 'operating on representations' alternative gives rise to the interpretation of representations as static data structures[17] awaiting some operation to be performed on them. On the other hand, the 'representation figuring in processes' alternative allows for representations to change dynamically. The former alternative is the basis for the classicist thesis, roughly saying that cognition is symbolic computation, where representations, which have a propositional format, are operated on by explicit language-like rules. Arguably, in connectionist networks distributed representations figure in activation-spreading processes and change dynamically.

However, in the context of accounts of digital computation proper, a further distinction should be drawn between *intrinsic* and *extrinsic representations*, which

---

[16] For a further discussion on connectionist networks see, for example, John Tienson's introduction (1988) or Másson & Wang (1990) for a more computer science-centric introduction.

[17] For Bill Bechtel (1998), static data structures used in symbolic computation do not simply refer to those that are allocated in memory statically, such as static arrays or static strings, but also to structures that grow dynamically in memory as required, such as linked lists and dynamic arrays. For instance, a static array is allocated in memory at a fixed size, as opposed to a dynamic array that is initially allocated in memory at a certain size, but its memory allocation size grows as it is populated with more and more elements.

are mutually exclusive.[18] We should distinguish what computer scientists call *formal* semantics from *real-world* semantics invoked by philosophers (White, 2011, p. 194). An *intrinsic* representation in a digital computing system is "confined" to the physical boundaries of that system and has some *formal* semantics, whilst an *extrinsic* one, which has *real-world* semantics, is not. Internal symbols, for example, are intrinsic representations, if their referents are also internal to the computing system. So, both the representer (e.g., a symbol or a string) and the representee (e.g., a memory register or an instruction) reside within the physical boundaries of the computing system. Internal reference to symbols and expressions in conventional digital computers is assigned and it is a primitive in the computer's architecture. Symbols in programming languages have formal semantics that is given by the semantics of those languages (White, 2011, p. 191).

The *semantics* of intrinsic representations is confined to the physical boundaries of the computing system. According to this semantics, the primitives of the computer language are interpreted as actions on the, somewhat abstracted, internal state of the computing system (White, 2011, p. 194). An example of an intrinsic representation is the primitive ADD operation, that is, arithmetic addition, in conventional digital computers.[19] It is described as a numerical opcode in the machine language, which is interpreted by the system as *standing in* for or *representing* the addition operation itself. This may invite the challenge that an interpretation by the computing system implies that it has knowledge-*that* of the instruction. But this is hardly the case in human-engineered computing systems.

A reply to this challenge requires a further distinction between knowledge-*how* and knowledge-*that*. Crudely put, the former is implicit knowledge, which is typically based on heuristics, whereas the latter is explicit and consists of propositional knowledge. This distinction is problematic though. Is the knowledge-how to do something amounts to *knowing* the relevant *facts* either implicitly or explicitly? Does knowledge-how consist in the knowledge that certain propositions *are* true? Some have argued that knowledge-how, such as how to ride a bike or how to play a piano, "cannot be analyzed in terms of abilities, dispositions and so on; rather, there appears to be an *irreducible cognitive element*" (Chomsky, 1992, p. 104, italics added). In other words, knowledge-how consists in knowledge-that (Stanley & Williamson, 2001). But, arguably, not all knowledge-how requires knowledge-that. "[I]f, for any operation to be intelligently executed, a prior theoretical operation [of considering appropriate propositions] had first to be performed [...], it would be a logical impossibility for anyone ever to break into the circle" (Ryle, 1949, p. 31).

---

[18] There is another type of representation that figures in computation proper, namely mathematical representation. It is a distinctive type of representation, which does not compete with either intrinsic or extrinsic representations. Mathematical representation is further discussed below.

[19] For an introduction to instruction sets in modern computer architecture, see, for example, (Stallings, 2010, Chapter 10).

Instead, knowledge-how is construed as a skilled performance of an operation that is measured by its success, efficiency, etc. (Ryle, 1949, p. 29). "Lessons impart information; ability is something else. Knowledge-that does not automatically provide know-how. [...] Know-how is ability. But of course some aspects of ability are in no sense knowledge" (Lewis, 2004, p. 100).

The plot thickens, but we need not go that far. To execute the primitive ADD operation, the CPU follows the opcode direction to the physical address of ADD. And the ADD operation itself is simply a hardwired mechanism[20] that converts input bits to output bits using some combination of logic gates. Put another way, the ADD operation is coded by a unique binary pattern and whenever this particular sequence lands in the CPU's instruction register, it is akin to a dialled telephone number that mechanically opens up the lines to the right special-purpose circuit (Dennett, 1991, p. 214). The CPU's "knowledge-how" requires no "knowledge-that" of the ADD operation.

Extrinsic representations, on the other hand, refer to symbols, data, events or objects that exist *outside* the physical boundaries of the computing system. Unlike their intrinsic counterparts, extrinsic representations are external-knower-dependent: a knower assigns external, or real-world, semantics to data structures, strings or symbols in the computing system. The contents of some data structures or computer programs may have external semantics relating to some states of affairs when the computing system directly interacts with the environment in which it is embedded. A particular database table, for example, could list names of employees and their respective salaries. The table, by convention, contains these data as bits and when the database is queried the records returned might contain strings of symbols representing the names of the employees and some numerals that represent the values of their salaries. Still, the relevant computer program will perform, say, the same database search operation, if prompted, just as well, even if the strings of symbols searched for were the names of planets, rather than of employees, and the corresponding numerals were their coordinates in the galaxy, rather than the employees' salaries.[21] External reference is not a primitive in computing systems.

---

[20] If, for some technical reasons, this mechanism is replaced with a soft-wired mechanism (i.e., either through explicit how-to rules or a soft-constraint learning mechanism), the overall principle still holds. Even in the case of the soft-constraint learning mechanism, the mechanism can eventually learn (say, by heuristics) *how* to perform effectively without knowing *what* precisely it does.

[21] At the program level, any user-entered information is converted into something recognisable by the computing system by using an implicit *semantics dictionary*. This dictionary is used to translate any user-entered information into some data structure that is recognisable by the program. An ace of hearts card, for instance, is translated into a data structure with attributed properties, such as a shape, a number and a colour. This data structure can be processed by the program and when appropriate, the processed data can be translated back into some form of human readable information as output.

The other type of representation that figures in computing systems is mathematico-functional.[22] An adder, for example, represents numbers – addends and sums. More generally, any system construed as computing a mathematical function represents the arguments and values of the function computed. Numbers, arguments and values of mathematical functions are neither intrinsic nor extrinsic to the computing system. For, as abstract objects, they do not exist in either space or time (regardless of whether or not they live in some Platonic realm). As suggested above, computation can be construed as the evaluation of a mathematical function on some input arguments in accordance with the classical theory of computability.

Mathematico-functional representations – or mathematical representations, for short – seem to still figure in computation even when any intrinsic representation is stripped off. If the physical properties of a concrete computing system are progressively abstracted away from, there remains a mathematical description of the function realised by that system (Scheutz, 1999, p. 171). This was already hinted above in discussing the extensional equivalence of the different formalisms of computability. When the particular physical details concerning inputs and outputs are abstracted away from, the mathematico-functional description of any two computing systems reveals what they both have in common, namely the *realisation of some function* (Scheutz, 1999, p. 174). Note that mathematical representation need not be limited to arithmetic functions on numbers. All that a mathematical function $f(x)$, *broadly construed*, requires is that f map a domain to a range and that the mapping of x be unique. A function whose domain and range are not numbers is still a function. Mathematical functions may also be defined on strings, sets, sequences and permutations.

Provided that this view is correct, then the distinction between intrinsic and extrinsic representations is not exhaustive and it has to be at least supplemented by mathematical representation. It seems clear that what gets computed in the course of a computation is the value of some specific input/output function. Accordingly, a characterisation of a system as computational amounts to its interpretation as representing the arguments and values of the mathematical function computed. Note that the same system could also be interpreted as representing some intrinsic properties, such as features of some data structure, or extrinsic properties, such as the spatial coordinates of some distal object.

From this it may be seen that it is extrinsic representation that resembles mental representation, at the appropriate level of abstraction. Mental representations, typically, stand in for some object, property or state of affairs in a world that is external to the cognitive agent. Similarly, extrinsic, but not intrinsic, representations stand in for things residing outside the physical boundaries of the computing system. The motivation behind the semantic view of computation is the

---

[22] This type of representation should not be confused with representation theorems in mathematics. Representation theorems concern identifying isomorphisms between some abstract structure (e.g., an algebraic structure) and a concrete structure. They are used to measure the generality arising out of abstraction (Joshi, 1989, p. 224).

following. If computational states and processes are individuated in the same manner that cognitive states and processes are, then computation is also individuated by extrinsic representations. Although this view has some obvious merits, it is highly contentious as we shall see in Chapter 5.

1.5.2   The Semantic View of Computation

This shared view can be best summarised in Jerry Fodor's words: "computation is a causal chain of computer states and the links in the chain are operations on semantically interpreted formulas in a machine code [...] There is no computation without representation" (1981, p. 122). Proponents of this view attribute semantic properties to computations that go on inside an artefact that is a computing system (Fodor, 1981; Pylyshyn, 1984, 1989; Shagrir, 2001, 2006; Smith, 1996, 2002, 2010).

Smith maintains that computation is a symbolic or a representation based phenomenon (Smith, 1996, pp. 9–11, 2002, pp. 32, 56, 2010, pp. 24–25, 34–35). On his view, the states of a computing system can model or *represent* other states in the world. Computation is an intentional phenomenon and that is partly why so many intentional words in English are used as technical terms in computer science, such as 'data', 'information' and 'representation'. This is analogous to the use in physics of terms such as 'energy' and 'force' that objectively represent states of affairs in the world. Smith argues that the intentionality of computation is the motivation behind the computationalist thesis. The only compelling reason to suppose that *cognition* could be explained *computationally* is that natural cognitive agents too, deal with representations, symbols and meanings. So not only do different states of a computing system represent 'something' for independent knowers, but they can also supposedly represent states of affairs, facts or whatnot in the absence of independent knowers.

In a similar vein, Shagrir maintains that content plays an individuative role in computation. By his lights, it is content that is used to determine which of the *various syntactic structures* simultaneously implemented by the computing system constitutes its *computational structure*, that is, the syntactic structure that defines the underlying computational task (Shagrir, 2001, pp. 370, 375, 2006). This is the source of the explanatory force of applying computational explanations to account for the way complex systems carry out semantic tasks. The crucial point here is identifying the correspondence relation between a computational structure of a system and certain semantic properties of the objects or states of affairs being *represented*. So, according to Shagrir, semantics plays an individuative role in terms of what is being represented by the corresponding computational structure. A computing system can implement different syntactic structures simultaneously. Thus, he concludes, in order to determine which syntactic structure *constitutes* the system's computational structure a *semantic constraint* has to be employed.

Another claim made by some proponents of this view is that symbolic expressions being manipulated in the process of computation maintain their semantic interpretation and keep their semantics coherent. Fodor, for one, asserts that

> "[i]t is feasible to think of [... a computational device] as a computer just insofar as it is possible to devise some mapping which pairs physical states of the device with formulae in a computing language in such fashion as to preserve desired semantic relations among the formulae. [... In] the case of real computers, if we get the right way of assigning formulae to the [computational] states it will be feasible to interpret the sequence of events that causes the output as a computational derivation of the output" (1975, pp. 73–74).

Zenon Pylyshyn argues that much of the computability theory is concerned with certain formal properties of symbol-manipulation system (1984, p. 57). The symbolic representations are manipulated in the course of computation, and the meaning of the symbols processed is coherently maintained by the *transformation rules*. These rules are meant to ensure that the generated symbolic expressions continue to make sense when semantically interpreted in a consistent way. The semantic properties of the manipulated symbolic expressions are preserved in a similar manner to truth preservation in logic. Yet, by his lights, if one wishes to ensure that the algorithm *always* gives the correct answer, one *has to* refer to facts and state of affairs, i.e., to the external *semantics* of the symbols (Pylyshyn, 1989, pp. 58–60).

In sum, according to the semantic view of computation, to paraphrase Egan, computational states proper inherently have their semantic properties (Egan, 1995, p. 194). On this view, resorting to computational explanation entails semantic content ascription to the explanandum (Piccinini, 2004, pp. 375–377). Proponents of this view aspire to ascribe semantic content to computing systems as a way of explaining cognition computationally. Accounts of computation that subscribe to this view are examined in Chapter 5. The motivation is clear: if both digital computation and cognition are semantically individuated, then proponents of the semantic view of computation expect that cognition can be explained computationally. But the question remains: are computational states individuated in the same way that cognitive states are?

1.5.3   The Causal View of Computation

Proponents of the causal view of computation offer an alternative analysis of computation according to which computation is individuated by its causal properties, rather than by any extrinsically representational properties (Chalmers, 1994; Copeland, 1996; Scheutz, 1999). Chalmers argues that the crucial bridge between abstract computations and physical computing systems is a theory of implementation (1994, pp. 391–393). For implementation is the relation that holds between the abstract computation and the physical system that executes the computation. Although some TMs, and their physical implementations, indeed yield to a systematic semantic interpretation, extrinsic representations are not part of the definition of either a TM or its implementation (Chalmers, 2011, p. 334).

Chalmers' analysis of implementation, underpinned by CSA[23] instead of the traditional FSA, assumes that a physical system P implements a computation if the causal structure of P mirrors the formal structure of the computation. Informally, P implements a given computation when there exists a grouping of physical states of P into state-types and a one-to-one mapping from formal computational states (i.e., the combinatorial states) to P's state-types, such that combinatorial states related by an *abstract* state-transition relation are mapped onto P's state-types related by a corresponding causal state-transition relation (Chalmers, 1994, p. 392). This characterisation introduces a bijection-relation requirement between physical states and computational states where the state-transitional structure of the physical computing system mirrors the state-transitional structure of the abstract computation. Hence, according to Chalmers' analysis, computation is individuated by its causal organisation: a computational description of a computing system is an abstract specification of its causal organisation (1994, p. 401).

A noteworthy causal account of concrete computation, which is offered by Copeland, is the algorithm execution account (1996). This account is examined in detail in Chapter 7. In the present context it is worth noting Copeland's notion of a labelling scheme introduced as a means for bridging the gap between the computing system's formal description and the computing system itself. "The trick", by his lights, is "to select states for labelling that *are* causally related in the required ways" (Copeland, 1996, p. 353).

Lastly, Scheutz argues that the standard concept of correspondence between physical states and formal computational states is inadequate and does not give a clear account of how computations mirror the causal structure of physical systems (Scheutz, 1999, 2001). Instead, he suggests that: "the state transitional structure of the computation must be at most as complex as the causal transitional structure of the implementing physical system with respect to computational sequences" (Scheutz, 2001, p. 560). Scheutz suggests that a better approach for understanding concrete computation is to start with physical descriptions of computing systems, and then gradually refine these descriptions, abstracting over some of the physical properties (1999, p. 192). This abstraction process stops when a unique computational description of the system is found. As a result, by his lights, the standard notions of computation and implementation are supplanted by the, supposedly, more viable alternative of "function realisation". The extension of the class of functions realised by digital computing systems then becomes empirically dependent on all *possible* physical theories of computation.

---

[23] A CSA is an automaton similar to an FSA except that all its states are described as vectors rather than monadic states. Each state of the CSA can be thought of as being composed of substates. State-transition rules are determined by specifying, for each substate of the state-vector, a function by which a successor state depends on the previous overall state-vector and input-vector (Chalmers, 1994, pp. 393–394). Chalmers' CSA formalism is further discussed in Chapter 4.

1.5.4 The Functional View of Computation

Another promising alternative to the semantic view of computation is the functional view of computation. Whilst the latter view is also causal in essence, it emphasises the importance of relevant *functional/organisational* properties of digital computing systems. Stephen Stich, for example, in arguing against the possibility of a scientific intentional psychology has implicitly rejected the semantic view of computation (1983, pp. 149–183). His main claim regarding folk psychology is that we need not refer to the *content* of cognitive states in our attempts to explain cognition. Cognitive states can be explained in terms of syntactic properties and relations of the abstract objects to which the cognitive states are mapped. As a consequence, this moral likewise applies to computational states, in the context of computational theories of cognition. On Stich's view, computational processes are sensitive to the syntax of the symbolic structures involved in these computations. Accordingly, an explanation of computation need not appeal to any semantic properties (Stich, 1983, pp. 192–193). Elsewhere, he has argued that "[o]n the computational view of mind, talk of representation is simply excess baggage" (Stich, 1980, p. 152).

Piccinini too claims that an appropriate kind of functional explanation is sufficient to individuate computation without appealing to extrinsic representations (2004, 2007, 2008). On his view, computational states are individuated by their functional/organisational properties. These functional properties are specified by a *mechanistic* explanation in a way that does not presuppose any semantic properties – or extrinsic representations, to use our present terminology. On his functional-mechanistic view, a mechanism has certain components, which have specific *functions*, such as the heart's blood-pumping function, and are *organised* in a specific way that gives rise to these functions. The mechanism's capacities are constituted by the relevant components, their functions and their particular organisation. To emphasise, these capacities are enabled by a special causal relation that obtains among these components in virtue of their *specific organisation*.

Similarly, MacLennnan characterises digital computation as a "physical process the purpose of which is the abstract manipulation of abstract objects" (2004, p. 121) without appealing to any extrinsic representations. On his view, computation is individuated by its teleological function (i.e., by *what* is done rather than *how* it is done). An *abstract object*, such as a number, a sequence, a set, a relation, a tree or a string, in his characterisation is one that is defined by its formal properties. *Abstract manipulation* includes the application of functions, such as exponentiation or positive square root, to their arguments and processes evolving in "abstract time" as in formal logic. The manipulation of objects, by his lights, is accomplished indirectly by means of physically manipulating physical representations, since abstract objects are not spatiotemporal entities. Accordingly, the bits of a conventional digital computer, for example, may represent numbers. However, this kind of representation, as in calculus, is merely formal – or syntactic (MacLennan, 2004, p. 117), since numbers, as remarked above, do not exist in either space or time. To use our present terminology, such representation is mathematical but not extrinsic.

1.6 The Plan of Attack

As mentioned above, the first objective of this book is to promote a clearer understanding of concrete computation, rather than just computability. To achieve this objective, some extant answers to the question "What does it take for a physical system to perform digital computation?" are examined. It should be noted that the key requirements for a physical system to perform computation are not to be taken strictly as both necessary and sufficient criteria for determining for *any* physical system whether it computes or not. There are bound to be some borderline cases, whose classification remains to be determined specifically by the relevant account. The second objective is to show that there is no compelling reason to posit that concrete computation is individuated by extrinsic representations.

In Chapter 2, we present and analyse the different criteria for evaluating the adequacy of extant accounts of concrete computation. This analysis is paramount, as it is the foundation for evaluating the various accounts on an equal basis. The concluding section of the second chapter lays out the recommended set of evaluation criteria. This set, so it is argued, includes all those criteria that are essential, and excludes those criteria that are over demanding.

Although the notion of computation preceded Turing's analysis of computability, the present enquiry starts with his account. This is the focus of Chapter 3. Despite Turing's analysis being grounded in idealised computing systems, rather than physical ones, it is of great importance to any venture that tackles the notion of concrete digital computation. Firstly, amongst the various formalisms of computability, it is the one most prominently invoked, for better or for worse, in cognitive science discourse in relation to computation. Secondly, it has been the source of many misconceptions of digital computation, including some attempts to trivialise this notion. Thirdly, all the accounts examined in this book fall back on TMs in some respect.

Chapter 4 examines the views of two well-known and influential philosophers concerning the allegedly ubiquitous character of concrete computation. The first part of the chapter is dedicated to explicating the reasoning behind Searle and Putnam's trivialisation of computation. Unsurprisingly, in the last part of that chapter, it is argued that their attempts can be blocked. That is fortunate, as otherwise there would have been little point reading beyond Chapter 4. This task is not overly complicated, since both Searle and Putnam have had their fair share of criticism in the 80s and 90s and even to this day. Still, understanding their reasoning is required for ensuring that adequate accounts of computation resist the trivialisation of computation.

Chapter 5 examines some semantic accounts of computation. The first one examined is the physical symbol system account. The second one examined is the formal symbol manipulation account. This latter account resembles the former in some respects, but it differs in the details and denotes a broader class of computing systems. The third one is the participatory account of computation that is reconstructed from Smith's views on computation.[24] Subsequently, we criticise some of the main arguments that proponents of the semantic view of computation have put forward in support of this view.

Next, Chapter 6 offers a comprehensive analysis of the information processing account. Although many cognitive scientists readily equate digital computation with information processing, it is hard to find an explicit account of computation in information processing terms. The second section of the sixth chapter introduces four different conceptions of information: Shannon information, algorithmic information, instructional information and factual information. In the subsequent two sections, we examine the features of the four resulting information-processing accounts, which are based on the conceptions above, and the key requirements they imply for a physical system to compute. The following section evaluates the adequacy of three resulting information-processing accounts that are based on Shannon information, algorithmic information and factual information. In the section that follows, we outline the instructional information processing account and conclude that it is evaluated positively against the recommended adequacy criteria. Some concluding remarks are given in the last section.

Chapter 7 analyses three non-semantic accounts of computation. The first group under analysis consists of Copeland's algorithm execution account and the Gandy-Sieg account, both of which are causal accounts of computation. The last one is the mechanistic account of computation, which emphasises the functional/organisational properties of computing systems. These three accounts share one important feature. They do not appeal to any extrinsic representations, and, thus, it is argued that, at least prima facie, they are preferable over accounts that do appeal to extrinsic representations. For our aim is to understand digital computing systems as *computational* systems and not as *representational* ones.

With a better handle on what it takes for a physical system to compute, in Chapter 8, we defend four theses concerning computation and computational cognitive science. In the first section, we argue that in the course of nontrivial computation, typically, only *implicit* intrinsic and mathematical representations are processed. In the second section, because the various accounts are both intensionally and extensionally distinct, thereby showing that 'digital computation' is an ambiguous term, we argue that any computational explanation of cognition is unintelligible without a commitment to a single interpretation of 'digital computation'. In light of this improved understanding of computation, the

---

[24] It is worth mentioning that this is not an account explicitly proposed by Smith. That would be contrary to his negative claim about the possibility of an adequate account of computation.

following section revisits the foundational role of computation in cognitive science. We argue that a blanket dismissal of the key role computation plays in cognitive science is unwarranted and also that the thesis that computationalism, connectionism and dynamicism are mutually exclusive is wrong. The book is concluded with some reflections on the computational nature of cognition.

References

Aizawa, K.: Computation in cognitive science: it is not all about Turing-equivalent computation. Studies in History and Philosophy of Science Part A 41(3), 227–236 (2010), doi:10.1016/j.shpsa.2010.07.013

Bechtel, W.: Representations and cognitive explanations: Assessing the dynamicist's challenge in cognitive science. Cognitive Science 22(3), 295–317 (1998), doi:10.1016/S0364-0213(99)80042-1

Bell, P., Staines, P.J., Michell, J.: Evaluating, doing and writing research in psychology: a step-by-step guide for students. SAGE Publications, London (2001)

Block, N.: Mental Pictures and Cognitive Science. The Philosophical Review 92(4), 499–541 (1983)

Boden, M.A.: Odd man out: Reply to reviewers. Artificial Intelligence 172(18), 1944–1964 (2008), doi:10.1016/j.artint.2008.07.003

Chalmers, D.J.: On implementing a computation. Minds and Machines 4(4), 391–402 (1994), doi:10.1007/BF00974166

Chalmers, D.J.: A computational foundation for the study of cognition. Journal of Cognitive Science 12(4), 323–357 (2011)

Chomsky, N.: Aspects of the theory of syntax. MIT Press, Cambridge (1969)

Chomsky, N.: Language and interpretation: philosophical reflections and empirical inquiry. In: Earman, J. (ed.) Inference, Explanation, and Other Frustrations: Essays in the Philosophy of Science, pp. 99–128. University of California Press, Berkeley (1992)

Cleeremans, A.: Computational correlates of consciousness. In: Laureys, S. (ed.) Progress in Brain Research, vol. 150, pp. 81–98. Elsevier (2005)

Copeland, B.J.: What is computation? Synthese 108(3), 335–359 (1996), doi:10.1007/BF00413693

Cummins, R., Poirier, P.: Representation and indication. In: Clapin, H., Staines, P.J., Slezak, P. (eds.) Representation in Mind: New Approaches to Mental Representation, pp. 21–40. Elsevier, Amsterdam (2004)

Decety, J., Lamm, C.: The Role of the Right Temporoparietal Junction in Social Interaction: How Low-Level Computational Processes Contribute to Meta-Cognition. The Neuroscientist 13(6), 580–593 (2007), doi:10.1177/1073858407304654

Dennett, D.C.: Consciousness explained. Penguin, London (1991),
http://www.contentreserve.com/TitleInfo.asp?ID=
23263F30-47C3-4241-B653-E61C277EECCA&Format=410 (retrieved)

Dennett, D.C.: Brainchildren: essays on designing minds. MIT Press, Cambridge (1998)

Dretske, F.I.: Knowledge & the flow of information. MIT Press, Cambridge (1981)

Dretske, F.I.: Explaining behavior: reasons in a world of causes. MIT Press, Cambridge (1988)

Egan, F.: Computation and Content. The Philosophical Review 104(2), 181–203 (1995), doi:10.2307/2185977

Egan, F.: Two kinds of representational contents for cognitive theorizing. Presented at the Philosophy and the Brain Conference 2011, The Institute for Advanced Studies, The Hebrew University of Jerusalem, Israel (2011), https://sites.google.com/site/philosophybrainias2011/home/conference-papers-1/Egan-TwoKindsofRepContent.pdf?attredirects=0&d=1 (retrieved)

Fodor, J.A.: The language of thought. Harvard University Press, Cambridge (1975)

Fodor, J.A.: Methodological solipsism considered as a research strategy in cognitive psychology. Behavioral and Brain Sciences 3(01), 63 (1980), doi:10.1017/S0140525X00001771

Fodor, J.A.: The mind-body problem. Scientific American 244, 114–125 (1981)

Fodor, J.A., Pylyshyn, Z.W.: Connectionism and cognitive architecture: A critical analysis. Cognition 28(1-2), 3–71 (1988), doi:10.1016/0010-0277(88)90031-5

Ford, K.W.: 101 quantum questions: what you need to know about the world you can't see. Harvard University Press, Cambridge (2011)

Fredkin, E.: Finite nature. In: Proceedings of the XXVIIth Rencontre de Moriond Series (1992)

Fresco, N., Staines, P.: A Revised Attack on Computational Ontology. Minds and Machines (forthcoming), doi: 10.1007/s11023-013-9327-1

Gandy, R.: Church's Thesis and Principles for Mechanisms. In: The Kleene Symposium, pp. 123–148. North-Holland (1980)

Gurevich, Y.: Interactive Algorithms 2005. In: Jedrzejowicz, J., Szepietowski, A. (eds.) MFCS 2005. LNCS, vol. 3618, pp. 26–38. Springer, Heidelberg (2005)

Harnish, R.M.: Minds, brains, computers: an historical introduction to the foundations of cognitive science. Blackwell Publishers, Malden (2002)

Haugeland, J.: Artificial intelligence: the very idea. MIT Press, Cambridge (1985)

Israel, D.: Reflections on Gödel's and Gandy's Reflections on Turing's Thesis. Minds and Machines 12(2), 181–201 (2002), doi:10.1023/A:1015634729532

Joshi, K.D.: Foundations of discrete mathematics. Wiley, Wiley Eastern Ltd., New York, New Delhi (1989)

Karp, R.M.: Reducibility Among Combinatorial Problems. In: Miller, R.E., Thatcher, J.W. (eds.) Proceedings of a Symposium on the Complexity of Computer Computations, The IBM Thomas J. Watson Research Center, Yorktown Heights, New York. Plenum Press, New York (1972)

Kellogg, R.T.: Fundamentals of cognitive psychology, 2nd edn. SAGE, Thousand Oaks (2012)

Kleene, S.C.: Mathematical logic, Dover (ed.). Dover Publications, Mineola (2002)

Lewis, D.: What experience teaches. In: Ludlow, P., Nagasawa, Y., Stoljar, D. (eds.) There's Something About Mary: Essays on Phenomenal Consciousness and Frank Jackson's Knowledge Argument, pp. 77–103. MIT Press, Cambridge (2004)

Lloyd, S.: Programming the universe: a quantum computer scientist takes on the cosmos. Knopf, New York (2006)

Longo, G.: Critique of computational reason in the natural sciences. In: Gelenbe, E., Kahane, J.-P. (eds.) Fundamental Concepts in Computer Science, vol. 3, pp. 43–70. Imperial College Press, London (2009)

MacLennan, B.J.: Field computation in natural and artificial intelligence. Information Sciences 119(1-2), 73–89 (1999), doi:10.1016/S0020-0255(99)00053-5

MacLennan, B.J.: Natural computation and non-Turing models of computation. Theoretical Computer Science 317(1-3), 115–145 (2004), doi:10.1016/j.tcs.2003.12.008

MacLennan, B.J.: Bodies — both informed and transformed embodied computation and information processing. In: Dodig-Crnkovic, G., Burgin, M. (eds.) Information and Computation, pp. 225–253. World Scientific (2011)

Másson, E., Wang, Y.-J.: Introduction to computation and learning in artificial neural networks. European Journal of Operational Research 47(1), 1–28 (1990), doi:10.1016/0377-2217(90)90085-P

Matthews, R.: Natural Computation. Presented at the Computation and the Brain Workshop 2011, The Institute for Advanced Studies, Hebrew University in Jerusalem, Israel (2011), https://sites.google.com/site/iascomputationbrainhuji 2011/home/previous-lectures/Matthews%282011%29Natural Computation.pdf?attredirects=0&d=1 (retrieved)

Miller, G.A.: The cognitive revolution: a historical perspective. Trends in Cognitive Sciences 7(3), 141–144 (2003), doi:10.1016/S1364-6613(03)00029-9

Minsky, M.L., Papert, S.: Artificial intelligence progress report. (AI memo No. 252), pp. 1–137 (1972), ftp://publications.ai.mit.edu/ai-publications/0-499/ AIM-252.ps (retrieved)

Newell, A.: Are there alternatives? In: Sieg, W. (ed.) Acting and Reflecting: the Interdisciplinary Turn in Philosophy. Kluwer Academic Dordrecht, Boston (1990)

Newell, A., Simon, H.A.: Computer science as empirical inquiry: symbols and search. Communications of the ACM 19(3), 113–126 (1976), doi:10.1145/360018.360022

Piccinini, G.: Functionalism, Computationalism, and Mental Contents. Canadian Journal of Philosophy 34(3), 375–410 (2004), doi:10.2307/40232223

Piccinini, G.: Computational explanation in neuroscience. Synthese 153(3), 343–353 (2006), doi:10.1007/s11229-006-9096-y

Piccinini, G.: Computing Mechanisms. Philosophy of Science 74(4), 501–526 (2007), doi:10.1086/522851

Piccinini, G.: Computation without Representation. Philosophical Studies 137(2), 205–241 (2008), doi:10.1007/s11098-005-5385-4

Piccinini, G., Scarantino, A.: Information processing, computation, and cognition. Journal of Biological Physics 37(1), 1–38 (2011)

Pinker, S.: So How Does the Mind Work? Mind and Language 20(1), 1–24 (2005), doi:10.1111/j.0268-1064.2005.00274.x

Pylyshyn, Z.W.: Computation and cognition: toward a foundation for cognitive science. The MIT Press, Cambridge (1984)

Pylyshyn, Z.W.: Computing in Cognitive Science. In: Posner, M.I. (ed.) Foundations of Cognitive Science, pp. 49–92. MIT Press, Cambridge (1989)

Pylyshyn, Z.W.: What's in your mind? In: LePore, E., Pylyshyn, Z.W. (eds.) What is Cognitive Science?, pp. 1–25. Blackwell, Malden (1999)

Ramsey, W.: Representation reconsidered. Cambridge University Press, Cambridge (2007)

Ryle, G.: The concept of mind. Penguin Books, Harmondsworth (1949)

Scheutz, M.: When Physical Systems Realize Functions... Minds and Machines 9(2), 161–196 (1999), doi:10.1023/A:1008364332419

Scheutz, M.: Computational versus Causal Complexity. Minds and Machines 11(4), 543–566 (2001), doi:10.1023/A:1011855915651

Schneider, S.: The Language of Thought: A New Philosophical Direction. MIT Press, Cambridge (2011)

Searle, J.R.: The rediscovery of the mind. MIT Press, Cambridge (1992)

Shagrir, O.: Content, Computation and Externalism. Mind 110(438), 369–400 (2001), doi:10.1093/mind/110.438.369

Shagrir, O.: Why we view the brain as a computer. Synthese 153(3), 393–416 (2006), doi:10.1007/s11229-006-9099-8

Shagrir, O.: Marr on Computational-Level Theories. Philosophy of Science 77(4), 477–500 (2010), doi:10.1086/656005

Shagrir, O.: Computation, Implementation, Cognition. Minds and Machines 22(2), 137–148 (2012), doi:10.1007/s11023-012-9280-4

Slezak, P.: The tripartite model of representation. Philosophical Psychology 15(3), 239–270 (2002), doi:10.1080/0951508021000006085

Smith, B.C.: On the origin of objects. MIT Press, Cambridge (1996)

Smith, B.C.: The foundations of computing. In: Scheutz, M. (ed.) Computationalism: New Directions, pp. 23–58. MIT Press, Cambridge (2002)

Smith, B.C.: Age of significance: Introduction (2010), http://www.ageofsignificance.org (retrieved)

Soare, R.I.: Turing oracle machines, online computing, and three displacements in computability theory. Annals of Pure and Applied Logic 160(3), 368–399 (2009), doi:10.1016/j.apal.2009.01.008

Stallings, W.: Computer organization and architecture: designing for performance, 8th edn. Prentice Hall, Upper Saddle River (2010)

Stanley, J., Williamson, T.: Knowing How. The Journal of Philosophy 98(8), 411–444 (2001)

Stich, S.P.: Computation without representation. Behavioral and Brain Sciences 3(01), 152 (1980), doi:10.1017/S0140525X00002272

Stich, S.P.: From folk psychology to cognitive science: the case against belief. MIT Press, Cambridge (1983)

Tienson, J.L.: An introduction to connectionism. The Southern Journal of Philosophy 26, 1–16 (1988), doi:10.1111/j.2041-6962.1988.tb00460.x

Van Rooij, I.: The tractable cognition thesis. Cognitive Science 32(6), 939–984 (2008), doi:10.1080/03640210801897856

Von Eckardt, B.: Multidisciplinarity and cognitive science. Cognitive Science 25(3), 453–470 (2001), doi:10.1016/S0364-0213(01)00043-X

Von Eckardt, B.: What is cognitive science? MIT Press, Cambridge (1993)

Wheeler, J.: The computer and the universe. International Journal of Theoretical Physics 21(6-7), 557–572 (1982), doi:10.1007/BF02650185

White, G.: Descartes Among the Robots. Minds and Machines 21(2), 179–202 (2011), doi:10.1007/s11023-011-9232-4

Wolfram, S.: A new kind of science. Wolfram Media, Champaign (2002)

Zuse, K.: Calculating Space. Massachusetts Institute of Technology, Project MAC (1970)

Chapter 2
An Analysis of the Adequacy Criteria for Evaluating Accounts of Computation

2.1 Introduction

In this chapter, we explore the possible criteria for evaluating the adequacy of candidate accounts of concrete digital computation. According to proponents of CTM, cognition is computational. Before we can judge the plausibility of any particular CTM, we need to understand what notion of computation this theory employs. Although there are extant accounts of computation, any of which may, in principle, serve as a basis for CTM, it is not clear that they are all equivalent or even adequate as accounts of computation proper. By proposing a plausible alternative to Smith's adequacy criteria, our goal here is to resist his discouraging claim that *no* adequate account of computation proper is possible.

An important disclaimer should be offered here first about 'account' versus 'theory'. The reader will have noticed a slippage between *theories* of computation and *accounts* of computation. Nothing significant hangs on this distinction for our purposes here. Crudely put, a scientific theory is a set of hypotheses about the relations that obtain among parts of the phenomenon in question as a whole. On an instrumentalist view, "[a] theory is a tool which we test by applying it, and which we judge as to its fitness by the results of its applications" (Popper, 2002a, p. 91). (For more details on philosophical analyses of explanation in scientific practice, see, e.g., (Hempel, 1965) & (Popper, 2002a).) Since some of the extant accounts, or *construals* as Smith calls them, of computation do not strictly advance falsifiable scientific hypotheses, we shall use 'accounts' to include more than theories, strictly. This choice allows the examination of the Triviality account in Chapter 4. Moreover, "computability theory", as it is referred to throughout the book, typically means the particular body of literature that rests on the definition of a TM.

Section 2.2 introduces the first set consisting of three adequacy criteria advocated by Smith. The first criterion states that an adequate account should be able to explain the computational systems that made Silicon Valley famous. The account has to distinguish *models* from *implementations* and *analyses* from *simulations*. The second criterion states that the account has to be conceptually

N. Fresco, *Physical Computation and Cognitive Science*,
Studies in Applied Philosophy, Epistemology and Rational Ethics 12,
DOI: 10.1007/978-3-642-41375-9_2, © Springer-Verlag Berlin Heidelberg 2014

rich for explaining all the underlying notions at the heart of computation, such as 'compilation', 'interpretation' and 'algorithm'. The third criterion states that the account has to support CTM, for computation has become an essential ingredient in cognitive science and philosophy of mind. By presenting some competing sets of adequacy criteria in the sections that follow, we show that there is a way around Smith's discouraging claim and argue that digital computation *can be* adequately explained.

Section 2.3 introduces the second set consisting of six adequacy criteria advocated by Piccinini. The first criterion states that an adequate account has to do empirical justice to theoretical computer science and computational practice. The second criterion states that the account has to identify computation objectively. The third criterion states that the account has to explain how program execution relates to the notion of computation simpliciter. The fourth criterion states that the account has to correctly distinguish the things that genuinely compute from those that do not. The fifth criterion states that the account has to explain the phenomenon of computational errors. The sixth criterion states that the account has to make possible a taxonomy of different types of computing systems.

Section 2.4 introduces the third set consisting of three adequacy criteria that were implicitly suggested by John von Neumann in an attempt to outline the logical foundations of computation simpliciter (1961). Von Neumann compared the logical aspects of computers with the central nervous systems of living organisms. The first criterion states that an adequate account has to extend beyond the limits of formal logic. The second criterion states that the account has to explain the possibility of miscomputation. The third criterion states that the account has to clearly distinguish *digital* from *analogue* computation.

Section 2.5 introduces the fourth set consisting of a single adequacy criterion, which is advocated primarily by Chalmers and Scheutz. On their view, a satisfactory account of computation is underpinned by an adequate theory of implementation. The single criterion states that an adequate account has to explain how abstract computation(s) are implemented by the physical computing system. A key motivation for this criterion is to resist the trivialisation of computation in physical systems.

In Section 2.6, a recommended set consisting of six adequacy criteria is proposed. This set offers a synthesis of those *justified* criteria that are required for evaluating the adequacy of accounts of concrete computation. It is argued that this set neither rules out *any* possible candidate account by setting the bar implausibly high nor does it admit *too many* accounts as adequate.

2.2 Smith's Tripartite Adequacy Criteria

According to Smith, an adequate account of computation has to meet the empirical criterion, the conceptual criterion and the cognitive criterion (1996, pp. 14–17, 2002, 2010, p. 14). He claims that the extant accounts fail to meet at least one of these criteria. On his view, the questions 'what are computers?' and 'what is computation?' require tackling questions of a metaphysical nature first. Any attempt to characterise computers as universal, programmable, or rule following systems inevitably appeals to higher order properties. But these properties are supposedly of the wrong metaphysical kind to be candidates for what is distinctive about computation proper.

Smith argues that not only must an adequate account meet the aforementioned criteria, but it must also include a theory of *semantics* and a theory of *ontology* (1996, pp. 69, 74, 2002, pp. 50–51). It is not just intentionality that is at stake, by his lights, but also metaphysics. The most serious problems that stand in our way of developing an adequate account of computation are as much ontological as they are semantical. For example, some may think that the formal symbol manipulation account (which is examined in Chapter 5) need not rely on any semantic foundations. However, at the ontological level, this account is *dependent* on semantics, viz., it is defined *in terms* of symbol *interpretation*.

2.2.1 The Empirical Criterion

This criterion states that an adequate account has to be compliant with the extant computational practices. This means that any candidate account should not only be sufficiently conceptually rich for explaining the operation of industry software programs, such as Open Office Writer, Firefox Web Browser, Google Maps and Mail clients, but also their *construction*, *maintenance* and everyday *use* (Smith, 1996, p. 5, 2002, p. 24). The motivation for the empirical criterion is keeping the analysis of concrete computation grounded in real-world examples of computing systems. The computer and Internet revolutions demonstrate time and again the computer's ability to evolve, expand and adjust beyond the alleged constraints of any particular account of computation. This criterion serves to question the legitimacy of extant theoretical perspectives. In this context, *Silicon Valley*, being a prototypical region of computer innovation, is nominated as a *gatekeeper* to determine whether, in practice, a given system should be deemed computational. An adequate account of computation has to make a substantive *empirical* claim about what Smith calls *computation in the wild*, which is the collective body of computational practices, techniques, networks, and so on (1996, pp. 5–6, 2002, pp. 24–25).

2.2.2 The Conceptual Criterion

This criterion states that an adequate account has to repay all the intellectual debts by explaining key concepts in theoretical computer science, such as 'compilation', 'interpretation', 'algorithm' and 'program' (Smith, 2002, p. 24). The conceptual criterion, as a meta-theoretical constraint, is especially crucial in the computational case for two main reasons. The first reason is that many accounts of computation rely on important notions, such as 'interpretation', 'representation' and 'semantics', without properly explaining these notions. The second is that there is a widespread tendency to appeal to the notion of computation for explaining those very "disobedient" notions. The end result is thus a conceptual circularity that deprives candidate accounts of their explanatory power (Smith, 2002, p. 24).

Smith argues that an adequate account of computation has to be either grounded in or at least defined in terms of a theory of semantics (2002, p. 33, 2010, p. 17, fn. 26). The semantic character of computation is revealed by commonly invoked phrases such as symbol manipulation, information processing, programming languages and databases. On his view, the conceptual criterion then implicitly requires that candidate accounts appeal to some theory of semantics to adequately explain the key concepts that underlie computation. But Smith does not stop there. He insists that candidate accounts also have to appeal to the relevant theory of ontology, for otherwise any attempt to explain computation will be liable to certain presuppositions, biases, and blindnesses regarding the phenomenon in question (Smith, 2002, p. 47).

By Smith's lights, accounts of computation inevitably have to deal with ontological quandaries (1996, pp. 27–28). These quandaries pertain to the very nature of computation. What are programs? What is the difference between programs and data structures? What is an implementation level and where does it border with abstraction? What is the ontological status of hardware and software? What is the ontological status of a virtual machine[1]? Is a computational state physical or abstract? What are the identity conditions on functions, algorithms and programs? And the list continues.

2.2.3 The Cognitive Criterion

This criterion states that an adequate account has to provide solid grounds for CTM (Smith, 2002, pp. 24–25). This criterion is also a meta-theoretical constraint on the *form*, rather than the substantive *content*, of any candidate account of computation. CTM have potential epistemological consequences depending on the account of computation proper one chooses to endorse. Any theoretical formulation of CTM is doubly contingent. Not only does a CTM make a claim about *natural cognitive agents* (e.g., people), for example, that these agents are formal symbol manipulators, but it does so by making a claim about *computing systems proper*, namely, that these systems are characterisable in the same manner.

---

[1] A virtual machine is a layer of software that is added to the "real" machine to support the desired virtualisation of its architecture. The virtual machine adds a level of abstraction for hiding lower levels of implementation. So, for example, the physical hard drive of a conventional digital computer may be presented as a set of files. For a nice exposition of this topic for a non-computer scientist, see, Sloman (2009)

2.2.4 Asking Too Much and Too Little

Smith rightly insists that candidate accounts do justice to computational practices or to what he calls *computation in the wild*. But elsewhere he claims that there is a big gap between the theory and the practice that the theory will *not* be able to overcome (Smith, forthcoming). Smith calls it *the explanatory gap*. A candidate account, by his lights, should not just do justice to computability and be detached from computation in the wild. Nevertheless, the technological gap, which is boosted by the computer and Internet revolutions, does not force us to give up the search for an adequate account in advance simply because it is supposedly left far behind. An adequate account need not explain every possible aspect all the technological breakthroughs in computer science, artificial intelligence and molecular computation. Instead, it needs to be sufficiently conceptually rich and explain the *fundamentals* of concrete computation.

Computation *simpliciter* is a highly diversified and fluid phenomenon, which may not be fully explained by a single unified account. A candidate account of *digital* computation should explain paradigmatic digital computing systems, including TMs, desktop computers, iPads, compilers, logic gates and CPUs. It can be limited to doing justice to paradigmatic digital computing systems only, but be *extended* to explain other complex systems, such as molecular computers and neural networks, provided indeed that they are justifiably classified as *digital computing* systems. The challenge clearly remains to ensure that candidate accounts not be so loose as to classify *any* system whatsoever as digitally computational, for this would result in pancomputationalism (see Chapter 4). Besides, on the view that an adequate theory is the one that is testable in the most rigorous way and has stood up to the severest tests (Popper, 2002a, p. 91), a previously adequate account of computation should simply be supplanted by another when it is falsified.

Smith claims that the extant accounts fail to meet the empirical criterion, for they are incapable of making sense of current computing systems and even less so when the "new generation" of systems is concerned (1996, pp. 8–9, 2002, pp. 28–31). His analysis leads him to conclude that it is not only that we *currently* do not have a satisfactory account of computation, but also that we will *always* fail to provide an adequate account (Smith, 2002, p. 51). This assertion is both too strong and unwarranted.

Smith's empirical criterion is hard to meet, but it does not justify dismissing the possibility of an adequate account of computation *in principle*. There is nothing preventing us from refuting a particular account of computation and providing a new one that addresses the weaknesses of its predecessor. In a true Popperian manner, a good scientific theory is subjected to falsification under the appropriate conditions (Popper, 2002a, pp. 9–10, 2002b, pp. 124–125). A scientific theory should undergo genuine tests in an attempt to refute it. This method of elimination promotes the survival of the fittest theories. These theories are tentative solutions to problems, which can *never* be *fully justified*. Similarly, when one account of computation is falsified, its weakness should be addressed by a new account that supplants it.

Smith's cognitive criterion is even more problematic than the empirical one. Whether computationalist claims are true or not need not be a meta-theoretical constraint on candidate accounts of computation proper. Regardless of the legitimacy of CTM, computationalist claims need not dictate the features of an adequate account of computation. If anything, it should rather be the other way around. The burden is on proponents of CTM to show why cognition works the same way that computing systems do. Smith rightly claims that if CTM were true, there would be significant epistemological consequences for the process of theorising itself. But if we take his view seriously, then CTM cannot be an adequate theory. To put it simply, according to CTM, cognition is a computing system. There will never be, by Smith's lights, an adequate theory of computing systems (2002, p. 51, 2010, p. 38). From these two premises it follows that there will never be an adequate CTM.

Smith also argues that if CTM were true, then an account of computation would apply not only to computation proper, but also to the process of theorising. In other words, any account's claims about the nature of computation would also apply to the account *itself* (i.e., as the product of the theorising process). So, if a particular CTM were true, then upon evaluating the adequacy of a candidate account of computation, there would be no reason (supposedly) to trust the result of the evaluation, unless this account is examined from a "reflexively consistent position" (Smith, 2002, p. 26). For the presumed meta-theory is conceptually inadequate.

The best approach to deal with Smith's argument is to simply avoid the trap. If CTM were false, then this criterion would become irrelevant anyway. However, if it were true, then the cognitive criterion would a-priori pre-empt any attempt to produce an account of computation. The burden of addressing the supposedly reflexive characteristic of the underlying theory lies on CTM, rather than on candidate accounts of computation proper. Computationalists maintain that cognition is a computing system and so they should address any reflexive implications of the underlying account of computation for CTM.

The conceptual criterion, despite appearing plausible at a first glance, also sets the bar too high for candidate accounts of computation. It requires that key underlying notions at the heart of digital computation be explicated by any candidate account. It is certainly reasonable to demand that concepts, such as 'algorithm', 'compiler', 'data' and 'program', be explained by any adequate account of computation. However, Smith also insists that any such account *be* at least *defined* in terms of a *theory of semantics*.

The difference between an algorithm and a program is known even if they are not defined in terms of a theory of semantics in textbooks in computer science. The semantics of key notions, such as 'program', 'algorithm', 'compiler' and 'virtual machine', is implicitly assumed by computer scientists and practitioners. Otherwise, it would be very hard to make any significant progress in computer technology, enable intra- and intercomputer communication as well as maintain consistency among different programming paradigms. Theories in physics are likewise not defined in terms of a theory of semantics for explicating key notions, such as 'energy', 'matter' and 'force'.

The demand that candidate accounts of computation include a theory of ontology is unwarranted. The ontological status of energy remains unsettled. Contemporary physical theories make use of the law of conservation of energy without having an ontological theory of energy. The ontological status of energy can be examined as a *consequence* of a particular physical theory. Similarly, the ontological status of programs is indeed an interesting topic for a philosophical enquiry. But why should it hinder the development of an adequate account of computation?

Metaphysics should not stand in the way of gaining better theoretical knowledge of concrete computation. A universal TM (UTM) nicely elucidates the distinction between a program and data without computability theory settling the ontological status of either programs or data. The operation of a MacBook computer can be explained without a complete explication of the ontological status of the *matter* from which it is built. Whilst the existence of programs, algorithms, data, and memory certainly has to be presupposed by any account of concrete computation, settling their ontological status should not be a prerequisite for any adequate account of computation.

It appears that Smith is asking *too much* but also *too little* with regard to potential accounts of computation proper. The foregoing criteria present insurmountable difficulties for any candidate account leaving very slim chances of providing an adequate one. On the other hand, he has overlooked other essential criteria, such as the miscomputation criterion and the *dichotomy* criterion. The former criterion states that any account of computation has to explain miscomputation as an inherent feature of *concrete* computation. It is discussed in more detail in the next section. The latter criterion states that paradigmatic computing systems have to be clearly distinguished from non-computing systems.

It might be objected that the dichotomy criterion is too strong, for digital computation is a graded concept. According to this line of argument, different systems range at different points on the scale. Some paradigmatic examples, such as TMs, conventional digital computers and iPads clearly compute and are located at one extreme end of the scale. Other examples, including digestive systems, walls and toasters do not perform (nontrivial) computation and are located at the opposite end of the scale. Whereas in the middle ground one may find systems, such as lookup tables, Ethernet cards and connectionist networks that may not always be clear-cut cases.

Nonetheless, an account of computation, which can explain a larger number of computing systems, is *ceteris paribus* better than one that caters for fewer systems. Noam Chomsky's analysis for determining which sequences of phonemes are "definitely [grammatical] sentences" and which "are definitely non-sentences" (2002, p. 14) can be adapted for the present purposes. An adequate account should include those systems that are "definitely computational" and exclude those systems that are "definitely non-computational". "In many intermediate cases we shall be prepared to let the [account] itself decide, when the [account] is set up in the simplest way so that it includes the clear [cases] and excludes the clear [non-cases]" (Chomsky, 2002, p. 14).[2] If an account of concrete computation achieves that, borderline cases shall be decided by the account. "A certain number of clear cases, then, will provide us with a criterion of adequacy for any particular [account]" (Chomsky, 2002, p. 14). This approach has a better chance of producing a broader account of concrete digital computation that is not too permissive.

Smith concludes that there is no distinct ontological category of computation, one that will be the *subject matter* of a deep and explanatory account (2002, pp. 50–51). "[C]omputing is instead a site: an historical occasion on which to see general, unrestricted issues of meaning and mechanism play out" (Smith, 2010, p. 40). The things that Silicon Valley calls computers do not form a coherent intellectually delimited class. By his lights, computers turn out to be analogous to cars. They are objects of personal, social and economical importance, but are not in themselves, the focus of an enduring intellectual enquiry. Computers are not "as philosophers would say, a natural kind" (Smith, 2002, p. 51). However, it is generally clear-cut whether an arbitrary object is deemed a car or a vehicle and what is involved in their operation. What qualifies as a computer, on the other hand, is not always as clear-cut, though any reasonable person would acknowledge that a personal desktop computer *is* a computer. It is not always so well understood what constitutes concrete computation as should become evident in the ensuing chapters.

Precisely because computation is not so well understood, an intellectual enquiry is called for. It may also be true that the things that *Silicon Valley* calls *computers* do not form a coherent intellectually delimited class. If that were the case, then the best thing to do would be to dismiss the requirement that something is a computer only if *Silicon Valley* so claims. Computation is a distinct subject matter and that is also the reason for the ongoing debate about computation and computers for nearly eight decades now. Computer science is the "living" proof of that.

Smith concentrates so much on "doing justice" to the practice and to CTM, that he neglects essential meta-theoretical constraints. Those "neglected" constraints are necessary to providing an adequate account of computation. The following sets of criteria illuminate what Smith has overlooked.

2.3 Piccinini's Sexpartite Adequacy Criteria

According to Piccinini, an adequate account of computation has to meet six criteria: the empirical criterion, the objectivity criterion, the explanation criterion,

---

[2] To be sure, Chomsky's analysis, strictly, deals with grammar and not with theories in general.

the right things compute criterion, the miscomputation criterion and the taxonomy criterion (2007, pp. 502–506). He maintains that computation has to be explained in *mechanistic* terms in a way that is analogous to engineering. Let us now examine the aforementioned criteria individually.

2.3.1 The Empirical Criterion

This criterion states that an adequate account should be grounded in existing real-world computing systems. It is only implicitly endorsed by Piccinini, whereas the remaining criteria discussed below are proposed explicitly. In a similar manner to Smith, Piccinini emphasises the importance of computational practice. However, the former questions the legitimacy of theoretical perspectives on computation and "nominates" Silicon Valley to decide whether a given system should be deemed computational. Smith's empirical criterion undermines the likelihood of ever producing an adequate account of computation. Piccinini, on the other hand, is implicitly committed to doing justice to the body of computational practices in a weaker sense.

2.3.2 The Objectivity Criterion

This criterion states that an adequate account has to identify computations as *a matter of fact*. Piccinini asserts that some philosophers (e.g., Searle and Putnam) have suggested that computational descriptions are vacuous, because any sufficiently complex physical system may be described as performing every digital computation (this view is the focus of Chapter 4). So, by their lights, there is no fact of the matter as to whether one computational description is more legitimate than another.

Yet, computer practitioners appeal to empirical *facts* about the systems they study, design and implement to determine which computations are performed by which systems. They apply computational descriptions to physical systems in a way analogous to other credible scientific descriptions (e.g., empirical descriptions used in physics to explain natural phenomena). Moreover, Piccinini argues that many psychologists and neuroscientists, who try to understand which computations the brain performs, appeal to empirical evidence about the systems they study (2007, pp. 502–504).

2.3.3 The Explanation Criterion

This criterion states that an adequate account has to explain how program execution relates to the notion of computation simpliciter. Computations performed by the system may produce some observable behaviour of that system. Typically, the behaviour exhibited by conventional digital computers is explained by appealing to the programs that they execute. CTM appeal to computations performed by cognition and frequently claim that cognitive processes can be explained in terms of *program execution* (Piccinini, 2007, p. 504).

However, Piccinini resists the reduction of digital computation to program execution. Whilst, traditionally, computational explanations have been underpinned by or reduced to explanations by program execution, he argues that there are examples of physical systems that compute, but not in virtue of program execution. For instance, by his lights, paradigmatic discrete neural networks *compute*, yet, they do not execute programs[3] (Piccinini, 2008a, p. 314). Conversely, there are systems that supposedly operate by executing "programs", but do not compute. Music boxes and automatic looms are mechanisms, which operate by executing programs, yet, they hardly qualify as digital computing systems proper (Piccinini, 2007, p. 517).

2.3.4 The Right Things Compute Criterion

This criterion states that an adequate account has to include *only* those systems that actually *compute*. Such an account should entail that paradigmatic examples, such as conventional digital computers, physical instantiations of TMs and FSA, as well as logic gates compute[4] (Piccinini, 2007, p. 504). On the other hand, it ought to *exclude* non-computing systems, such as planetary systems, digestive systems, Hinck's pail[5] and Searle's wall. Conventional digital computers, calculators, TMs and FSA compute and constitute the *subject matter* of computer science. To the extent that the working assumptions of computer scientists ground the success of computer science, these assumptions ought to be respected (Piccinini, 2007, pp. 504–505). This criterion is also meant to resist the trivialisation of computation, which implies that there is no distinction to be drawn between genuine computing systems and non-computing systems.

2.3.5 The Miscomputation Criterion

This criterion states that an adequate account has to address the fact that a physical computing system may miscompute, i.e., the computation may go wrong. A system S is said to miscompute, just in case S computes a function F on input I,

---

[3] If we granted that discrete neural networks *do* digitally compute, it would certainly not be clear that they do so *by* executing programs. For they lack a central control unit (which exists in UTMs and in any modern digital computer) and the capacity to store multiple instructions in each individual neural unit.

[4] Clearly, some physical systems, such as logic gates, perform only trivial computations, whereas physical instances of special-purpose TMs perform nontrivial computation. This point is further discussed in Section 2.3.6.

[5] Ian Hinckfuss presented the problem case known as 'Hinck's pail' to attack the functionalist theory of mind in a discussion at the 1978 AAP conference in Canberra. He described a pail of spring water in which at the micro level a vast complexity of things is going on. At the molecular level, an even more complex activity is required to sustain the micro level 'things' in the water. Some may argue that this underlying complex activity might realise a *program* for a brief period (Copeland, 1996, p. 336).

where $F(I) = O_1$, but S outputs O_2, where $O_2 \neq O_1$. O_1 and O_2 stand for any possible outcome of a computation, including the possibility that F is undefined for a particular I. An adequate account of computation should explain how it is possible for a physical system to miscompute. Miscomputation plays an important role in computer science and in "computation in the wild". Computer practitioners devote a large portion of their time and efforts to both avoiding miscomputations and coming up with appropriate techniques to find them and deal with them in the course of executing the computation (Piccinini, 2007, p. 505).

2.3.6 The Taxonomy Criterion

This criterion states that an adequate account has to distinguish among the capacities of different classes of digital computing systems. For instance, logic gates, which are very basic components in conventional digital computers, perform *trivial* computations on single bits (a NOT gate) or pairs of bits (e.g., a 2-input, 1-output OR gate). More complicated systems, such as nonprogrammable calculators can compute a finite number of functions for inputs of bounded size. Modern digital computers are programmable and can, in principle, compute any recursive function on input of any size subject to a constant supply of energy until they run out of either memory or time. An adequate account of computation should cater for a taxonomy of computing systems based on their computational capacities (Piccinini, 2007, p. 505).

An account of computation that cannot classify the computational capacities of extant computing systems is *ceteris paribus* less preferable than one that can. For instance, Robert Cummins' account of computation cannot distinguish between computing systems of different computational powers. On his account, computing systems compute in virtue of *program execution*. But then we can hardly distinguish the computational capacities of UTMs from special-purpose TMs, FSA and calculators. For, by his lights, either both FSA and special-purpose TMs compute in virtue of program execution or they do not. However, suppose we granted that special-purpose TMs execute programs, it would still be hard to explain why they are less powerful than conventional digital computers. The difference between computing systems that execute programs and those that do not is important to computer science and it should also make a difference to CTM (Piccinini, 2007, p. 506).

2.3.7 An Adequate Alternative

Although some of Piccinini's criteria require some fine-tuning, they serve as an adequate alternative to those propounded by Smith. The former presents decisive criteria that serve to discriminate between a satisfactory account of computation and an unsatisfactory one. Piccinini's proposed criteria demand that adequate accounts explain essential characteristics of digital computation, for example, the possibility of miscomputation and the computational power each system has.

The objectivity criterion states that whether or not a physical system computes is a *matter of fact*. This criterion is certainly not problem-free. By Piccinini's lights, performing digital computation is an empirical fact similar to the functional role of the heart to pump blood or the photosynthesis process in plants to produce glucose. Scientists from various disciplines resort to empirical studies to explain different phenomena and so do computer scientists when dealing with computations. Accordingly, an adequate account of computation ought to provide a suitable explanatory framework in which attributing digital computation to some system is not done arbitrarily.

It might be objected that the requirement for the objectivity of computation is overrated. According to this objection, the same computing system may be given *different* descriptions according to which a specific operation of that system is attributed two or more distinct computations. This possibility supposedly suggests that computation is knower-relative[6], rather than genuinely inherent to the system in question, thereby rendering the objectivity criterion irrelevant. For instance, a basic AND gate (under its conventional description) could just as well be described differently as an OR gate (Bishop, 2009, p. 228; Shagrir, 2001, p. 374). Similarly, some more complex systems, say, originally programmed to play chess, under the right conditions, produce unintended geometrical designs on computer screens. Although these systems were programmed to operate on symbols having nothing to do with visual figures, when their internal storage cells were interpreted as numbers representing points on a display screen, "strikingly regular [visual] patterns emerged" (Winograd & Flores, 1986, p. 92).

However, it does not follow from these examples that digital computation is merely knower-relative. The issue of knower-relativity is discussed in more detail in Chapter 4 in reply to the Triviality Thesis. But for now, let us make a few brief observations. Even if we interpreted a particular AND gate G in two different ways (say, as an AND gate and as an OR gate), we would still need to follow the "rules of the game". The two "interpreted" versions of the same gate could only co-exist in a single digital circuit, if we treated them as such, say, by adding some buffer between the two versions of G to make the traditional logic (G as an AND gate) and negative logic (G as an OR gate) compatible[7]. Besides, it is easier to assign computational descriptions to stateless trivial computing systems than to nontrivial ones. G can indeed be interpreted as either an AND gate or an OR gate. This interpretation is possible, because $\neg(p \wedge q)$ is equivalent to $\neg p \vee \neg q$ allowing the input and output signals of G to be interpreted in either classical logic (yielding logical conjunction) or negative logic (yielding logical disjunction).

---

[6] Typically, critics of the objectivity of digital computation characterise it as an *observer-relative* phenomenon. However, by replacing the *observer* with a *knower* some potential epistemic implications can be avoided. Rather than committing to a purely empiricist view, the knowledge of a knower could be based as well on a priori reasoning, and/or some other epistemic source.

[7] This claim is in accord with comments by Moshe Bensal (personal communication).

Computational descriptions cannot be assigned arbitrarily. It is not sufficient that only the initial state(s), input(s) and output(s) be given some coherent semantic interpretation. This semantic interpretation has to be such that all the state transitions of the system (in accordance with the appropriate computability formalism) are accounted for in a coherent manner. As the design complexity of the computing system increases (e.g., a UTM versus a stateless basic logic gate), the number of *consistent* computational descriptions that can be assigned to it decreases. The physical state to abstract computational state mapping will always determine the function that the particular system implements.[8] There indeed "remains the logical possibility that a computer could end up operating successfully within a domain totally unintended by its designers"; still, this is only "accidental" (Winograd & Flores, 1986, p. 92).

It is likewise crucial for any candidate account to show why certain systems perform computations, whereas others do not. Some philosophers (most notably Searle and Putnam) assert that any sufficiently complex physical system can be deemed computational. Such views trivialise the notion of digital computation, thereby undermining the need for accounts of computation. If everything can be deemed a digital computing system, then computational explanations become implausibly weak and lose much philosophical interest. Such explanations are inconsistent with those employed in both theoretical computer science and computational practice. For a computational explanation to be properly applicable to physical systems, it must respect the causal structure of the particular system. When applied correctly, not every sufficiently complex physical system can be classified as computational (Chalmers, 1996; Chrisley, 1994; Copeland, 1996; Piccinini, 2008b). Searle and Putnam might, of course, claim that their nonstandard interpretations do respect the causal structure of the computational systems in question. Their nonstandard interpretations are criticised in the last section of Chapter 4 and Section 7.2 of Chapter 7.

The miscomputation criterion is yet another important feature of any adequate account of computation. Unlike idealised TMs, physical computing systems are susceptible to physical noise that may result in an abnormal behaviour. Physical noise is any type of interference introduced in the physical environment of the system that impedes the normal process of communication or computation. It is sometimes referred to as "external noise", because the factors interfering with the system concerned are *external* to the system. The abnormal behaviour of the computing system may typically end in one of two ways. On the one hand, the system may handle the isolated error and resume its otherwise normal functioning (whilst possibly skipping some steps and/or losing some output in the process). On the other hand, it may either stop functioning completely (e.g., terminate unexpectedly) or produce a wrong output. This type of miscomputation is known

---

[8] This point is acknowledged by Mark Bishop, who advocates the knower-relativity of computation (2009, p. 228, fn. 7). Yet, the individuation of the appropriate physical "states" is not an easy task as is discussed further in Section 2.5.1.

in computer science as a *fault* or *malfunction*. Although there is sometimes the expectation that software (as well as hardware) be *bug free*, this is rarely ever the case.[9]

A malfunction in conventional computing systems is an error in the computation process due to either a hardware malfunction or certain types of runtime error in software. A hardware malfunction could range from any single component of the computing system (e.g., a memory register, a CPU, a data path or an I/O device) to multiple components breaking simultaneously. This malfunction could be the result of normal wear and tear of the physical components, exposure to extreme temperatures and/or dust, irregular spikes of electricity or in general physical noise. It can also be the result of the breakdown of a non-computing system, such as the cooling system. Certain runtime errors induced by software may also result in a malfunction. For example, a faulty software design can incorrectly lead to a constant voltage feed in some chip that results in the chip's meltdown. For a detailed analysis of computational errors in conventional digital computing systems and their taxonomy, the reader is referred to (Fresco & Primiero, 2013).

It should be noted that Piccinini's definition of a miscomputation introduced in Section 2.3.5 cannot be correct for all concrete computations. In Chapter 5, we consider an example of a computational process analogous to a two-step deductive argument (P→R→Q), where the second step (R→Q) is *invalid* and yet the overall argument (P→Q) is *valid*. Although the final outcome of the computation is the same as the intended one (when the system functions properly), we would still say that the system concerned miscomputed. For a computational system may produce the correct output (analogous to Q in our case) fortuitously as a result of some physical noise. The computing system may still produce the same output that would have been produced had it not malfunctioned. If it does so by following an *incorrect* computational step (analogous to R→Q in our case), then it still miscomputes.

Arguably, miscomputations are analogous to misrepresentations. This claim presupposes that computation resembles representation to start with. To reiterate the point from Chapter 1, according to Fred Dretske, only if a system has the capacity to misrepresent, does it have the power to get things right, in a way approximating *meaning*. But on this account of misrepresentation, conventional digital computers only derive *the capacity to represent or misrepresent* from their human designers or users, who already have the full range of intentionality. The computer's capacities to represent are merely reflections of *our* minds.

Despite the similarity between miscomputation and misrepresentation, they are *not the same*. Clearly, any coherent account that *equates* the former with the latter is unavoidably committed to digital computation being inherently semantic.

---

[9] This is not to say that every program inevitably contains bugs, but it rather refers to the more complex programs. Clearly, a trivial program comprised of a single step, which prints '*Hello World*' as output, would most likely be *bug free*.

This is an implication of the distinction introduced in Chapter 1 between intrinsic and extrinsic representations. A digital computing system may compute correctly or incorrectly regardless of whether it *represents* or *misrepresents* anything extrinsically. But when the opcode for SUB is treated as ADD, for example, a miscomputation is guaranteed. In this case, a mathematical misrepresentation occurs, namely, the addition function is incorrectly invoked instead of subtraction. The algorithm's instruction that incorrectly triggered the ADD opcode expecting the function's return value to be that of SUB would result in a miscomputation. This is also a case of intrinsic misrepresentation where a string (SUB) is interpreted as the wrong instruction to place the sum of two addends in the result field.

But miscomputations may occur regardless of whether the (correct) computation process represents anything external to the computing system or not. A computing system may miscompute in a variety of ways, such as running out of memory or some hardware malfunction. Some miscomputations may even result in a complete halt of the computation process, in which case the end result is not *incorrect* but simply *non-existent*. And clearly, *ontic nothing* can neither easily represent nor misrepresent anything.

Certainly, much more can be said about miscomputation, but to the extent that this is a pervasive phenomenon in computing systems, an adequate account of computation should explain it. Miscomputations are likely to manifest themselves in most practical applications and physical computing systems. Computer practitioners devote so much time attempting to handle as many potential miscomputations as practical. So, miscomputation should not be ignored by adequate accounts of computation.

Lastly, the taxonomy criterion proposed by Piccinini assumes the truth of the Church-Turing Thesis, which states that every effectively computable function is computable by a TM. As discussed in the next chapter, Turing introduced the notion of a UTM for a machine that can simulate any ordinary TM, thereby computing any recursive (or Turing-computable) function. The extant formalisms of computability (e.g., Gödel's recursive functions model and Church's lambda calculus) are Turing-equivalent. Whether this extensional equivalence suffices for computation simpliciter remains contentious. Any problem that cannot be solved by a TM, given the Church–Turing Thesis, presents a limit to what can be accomplished by any form of machine that works in accordance with effective methods. It remains an open empirical question whether there are in fact deterministic physical processes that cannot be simulated by a TM (Copeland, 2004, p. 15).

In sum, Piccinini's set of criteria improves on Smith's proposed criteria and paves the path for theorists of computation. The *objectivity* criterion and *the right things compute* criterion essentially have the same motivation, namely that computing systems be distinguished from non-computing systems. Otherwise, it is a slippery slope to describing any sufficiently complex physical system as computational. The *explanation* criterion[10] amounts to clarifying the relation between program execution and digital computation and whether the latter is always reducible to the former.

2.4 Von Neumann's Tripartite Adequacy Criteria

Back in the late 40s von Neumann claimed that we were very far from possessing a proper logical – mathematical theory of automata (1961). He argued that formal logic deals with rigid, all-or-none concepts and has very little bearing on the continuous concepts of the real and complex numbers. His motivation for announcing the need for such a theory was the unlikelihood of constructing automata of a much higher complexity than the ones, which existed then, without it. Von Neumann argued that dealing with high reliabilities of and error checking in high-speed computing systems is crucial. Any exhaustive study and a nontrivial account of computation must take them into account. Accordingly, von Neumann promoted the precision and reliability criterion as well as the single error criterion, but also the analogue – digital distinction criterion for candidate accounts of computation *simpliciter*.

2.4.1 The Precision and Reliability Criterion

The results of complex computations performed by computing systems may depend on a sequence of a billion steps and have the characteristic that every step actually matters or, at least, may matter with a considerable probability. This is the most specific and most difficult characteristic of computing systems (Von Neumann, 1961, pp. 291–292). In dealing with modern logic, the important thing is whether or not a result can be achieved in a finite number of elementary steps. The *number* of steps, which are required, is hardly ever a concern. In formal logic, any finite sequence of correct steps is, as a matter of principle, as good as any other. On the other hand, whilst digital computing systems obey the rules of formal logic, what matters, is not just whether the system can reach a certain result in a finite number of steps, but also how many such steps are required. In computer science, the time complexity of a problem is typically measured by the number of steps it takes to solve an instance of that problem as a function of the size of the input.[11] The criterion that may be distilled from this analysis, then, is that an adequate account of digital computation has to extend beyond formal logic.

---

[10] It seems odd to call a criterion for evaluating the adequacy of accounts or theories of some phenomenon 'an explanation' criterion. Instead, in what follows, this criterion is referred to as the *program execution* criterion.

[11] To avoid the need to analyse the runtime complexity of an algorithm for *every type* of computer, the number of steps required by the algorithm to solve the problem is generally used. For it typically does not vary much between computers.

Moreover, there are a couple of reasons for the key difference between formal logic and digital computation. Firstly, physical computing systems are constructed in order to reach certain results within pre-assigned durations (as opposed to, say, idealised TMs, which can use unbounded time to solve a particular problem). Secondly, the componentry employed in physical computing systems has a non-zero probability of malfunctioning in every instance of its operation (Von Neumann, 1961, pp. 303–304). The reader is reminded that idealised TMs, unlike physical computing systems, are *not* susceptible to malfunction.

2.4.2 The Single Error Criterion

Von Neumann compared error handling of physical computing systems to that of living organisms. He asserted that the organism itself, without any significant external intervention, corrects malfunctions, which occur in it. The organism must, therefore, contain the necessary arrangements to diagnose errors, as they occur, in order to minimise their effects and to correct or block the component at fault. Error handling in computing systems, on the other hand, is treated differently. In practice, every effort is made to detect any error as soon as it occurs. An attempt is then made to isolate the erroneous component as fast as practical. The basic principle of *nature* in dealing with errors is to make their effect as harmless as possible and to apply correctives, if required. However, in *computing systems* an immediate diagnosis is required, therefore, any attempt is made to ensure that errors are as conspicuous as possible. In this way, error correction methods can be applied immediately after diagnosis.

A computing system could be designed so that it operates almost normally in spite of a limited number of errors. But as soon as the system has begun to malfunction it would most likely go from bad to worse and only rarely fully recover from the error. So, it is essential that actions to correct errors be taken immediately. According to von Neumann, the error-diagnosing techniques used at the time were based on the assumption that the computing system contained only one faulty component. When this was the case, an iterative subdivision of the system into its subcomponents was used to determine which portion contained the fault (Von Neumann, 1961, pp. 305–306). The criterion that may be distilled from these observations is that an adequate account of computation has to explain the possibility of miscomputation.

2.4.3 The Analogue – Digital Distinction Criterion

All physical computing systems fall into two main classes in an obvious way. This classification is into analogue and digital systems.[12] Von Neumann presupposed

---

[12] Von Neumann refers to analogue computers as *analogy machines* or *analogy automata*. Here, the more common term 'analogue' is used instead to avoid the debate about the analogue – analogy comparison.

that computation is a numerical operation. An analogue computing system is based on the principle that numbers are represented by continuous physical quantities. Such quantities might be, for instance, the amperage of an electrical current or the size of an electrical potential. An analogue device, for instance, may take two currents as its input, so that its output current has a magnitude equal to the product of the input currents' magnitudes. Analogue computers have inherently limited accuracy due to their exposure to physical noise. The guiding principle concerning this type of computation, according to von Neumann, is the *signal to noise ratio*. Hence, the question can be formulated as follows. How large are the uncontrollable fluctuations of the system (i.e., physical noise) compared to the significant signals that express the numbers on which the system operates? The critical problem of any analogue computer is how low it can keep the relative size of the uncontrollable noise level (Von Neumann, 1961, pp. 292–293).

According to von Neumann, a digital computing system is based on the method of representing numbers as sequences of digits. Digital computers represent quantities by discrete states, operate serially and have unlimited accuracy in principle. The basic operations of a digital computing system are typically the four main arithmetical operations, namely, addition, subtraction, multiplication and division. Prima facie, one might mistakenly think that a digital computing system has absolute precision. However, even if the operation of each component produces only fluctuations within its pre-assigned tolerance limits, errors eventually creep in.

Importantly, the difference between digital and analogue computing systems lies in their ability to reduce the fluctuations. Noise level can be reduced in digital computing systems in an increasingly easy manner compared to analogue systems (Von Neumann, 1961, pp. 295–296). Consider, a flip-flop used in conventional digital computers that operates, say, within the voltage range [0v – 0.8v] and [1.7v – 2.5v]. It distinguishes between a logical 0 and a logical 1 by using a threshold. Fluctuations in the voltage, say +0.2v or -0.2 volts do not adversely impact the value the flip-flop settles on. On the other hand, consider an integrator circuit – an important component in analogue computers – that produces as output voltage the integration of the input voltage. Every integrator can introduce some amount of drift due to leakage in its integration capacitor (Ulmann, 2013, p. 43), thereby adding a degree of instability to the computer. The criterion that may be distilled in the present context is that an adequate account of computation *simpliciter* has to clearly distinguish digital from analogue computing systems.

2.4.4 A Comparison with Previous Criteria

Von Neumann promoted the three criteria above in the late 40s coinciding with the design and construction of the first physical digital computers (notably, the 1943 British Colossus and the 1945 American ENIAC). A full-blown account of concrete digital computation was not yet available at that time; the focus then was rather on effective computability, most notably by Turing, Church and Post, independently. Still, von Neumann's proposed criteria remain largely relevant today. The distinction he drew between formal logic systems and physical computing systems as regards the importance of the number of steps needed to solve a computational problem lies at the heart of theoretical computer science under the heading *complexity theory*.

Furthermore, the first and second criterion emphasise the significance of miscomputation being addressed by adequate accounts of digital computation. Whilst there seems to be no reason to insist on the single error principle, it is certainly important to cater for the fact that physical computing systems are susceptible to physical noise to a varying degree. The exposure to noise and the possibility of miscomputation exist both at the software and the hardware levels of computing systems.

Lastly, the distinction between analogue and digital computation is important even today when there is still no obvious consensus on what computation simpliciter is. Although it seems straightforward, an adequate account of concrete digital computation should clearly distinguish digital from analogue computation.[13]

2.5 Implementation Theory Bridging Computability and Concrete Computation

To resist the trivialisation of digital computation, according to Scheutz (1999) and Chalmers (1996), a satisfactory account of computation should be underpinned by an adequate theory of implementation. The former asserts that the notion of implementation should not be construed as the standard physical state to computational state correspondence, but rather as realisation of functions (Scheutz, 1999). The latter argues that a better understanding of the bridge between computability and concrete computation can be achieved in terms of implementation of CSA (Chalmers, 1996).

2.5.1 The Implementation Adequacy Criterion

This criterion states that an adequate account has to either rely on an existing or provide a new satisfactory theory of implementation. According to Putnam's realisation theorem every ordinary open system is a realisation of every abstract FSA (1988, pp. 121–125). For the purposes of the present discussion, let us assume that by "every ordinary open system", Putnam essentially means every physical object (this assumption is further examined in Chapter 4). On this view, every physical object, including a rock or a chair, can be viewed as implementing *every* program (Chalmers, 1996). Scheutz argues that Putnam's theorem shows that any account of computation lacking an adequate theory of implementation is built on weak grounds (Scheutz, 1999, p. 162).

---

[13] The reader is encouraged to remember this point in the context of discussing connectionism in Chapter 8.

Scheutz suggests that digital computation should be dealt with from a practical point of view by looking at existing applications and systems that are designed, implemented and used by people (1999, pp. 162–163). Rather than asking how abstract computations relate to physical systems, we should start with physical computing systems to determine how they are different from non-computing systems. On his view, this approach results in a more restricted notion of function realisation that is based on physical constraints (e.g. measurability, feasibility, and error range).

Importantly, one of the defining characteristics of the standard notion of digital computation is its independence from the physical substrate that realises it (viz., multiple realisability). This means that the same abstract computation can be executed on a range of different physical systems. Concrete computing systems are physically situated in the world and by simply appealing to the abstract levels, we tend to lose sight of this fact. Unlike in the case of formalisms of computability, accounts of concrete computation have to explain how abstract computations are implemented by physical systems. By Scheutz's lights, a theory of implementation should not depend on the standard physical-to-computational state correspondence, but rather exploit descriptions of certain properties of physical systems and abstract computations (1999, p. 162). It should appeal to the link between the concrete and the abstract, while emphasising the physical constraints of the implementing systems.

More specifically, singling out the relevant physical states that are mapped to the computational states is problematic. The physical states of a given object, O, are typically defined by the physical theory in which that object is described. But O might have too many physical states that possibly correspond to some computational state. In order to exclude unwanted states of O from consideration, an individuation criterion is required for picking only the right physical states. However, this criterion is not defined within the physical theory that is used to describe O, but rather at a higher level of abstraction. Still, that level of abstraction must be lower than the computational one to avoid circularity (Scheutz, 1999, p. 168).

2.5.2 The Empirical Criterion and the Cognitive Criterion Revisited

Scheutz's *approach* to implementation implies a weaker version of Smith's empirical criterion. The former maintains that his approach honours computational practice. Computer practitioners define computation in terms of the functions realised by concrete systems, rather then by an appeal to abstract computations independently of their realisation. The latter concludes that his empirical criterion precludes any satisfactory account of digital computation. In contrast, Scheutz implicitly invokes the empirical criterion in a constructive manner when theorising about implementation, which clearly has consequences for subsequent accounts of computation.

Scheutz also acknowledges that the notion of implementation used for explicating concrete computation has unavoidable implications for CTM (1999, pp. 191–192), in a like manner to Smith's justification for the cognitive criterion. By Scheutz's lights, if cognition can be adequately explained at the same level at which digital computation proper is explained, then CTM is true. However, if it can only be adequately explained at a level lower than the "digital one", then either the notion of computation has to be changed to apply to the class of functions realised by systems described at that level (thereby rendering CTM true) or CTM is false.

2.6 Recommended Sexpartite Adequacy Criteria

What then should be the minimal set of criteria for evaluating the adequacy of candidate accounts of concrete digital computation? Smith claims that digital computation is inherently intentional, thereby prompting him to propose the *cognitive criterion*. He maintains that the misconception of computation as being entirely abstract and formal hides the true representational character of computation. Thus, he argues that any adequate account of computation has to be founded on a sound *theory of semantics*.

Yet, the cognitive criterion introduces an unnecessary circularity. Theorists of computation supposedly need to consider the potential consequences of candidate accounts of computation for CTM. Interestingly, when considering this criterion in conjunction with Smith's claim that there will never be an adequate account of computation, the result is curious. The computationalist claim that cognition is computational combined with Smith's negative claim above entail that there will never be an adequate CTM (when tacitly assuming that CTM is essentially underpinned by a theory of digital computation proper). This seems to defeat the purpose of introducing the cognitive criterion in the first place.

Indeed, digital computationalism should be reflected upon in light of our best candidate account of digital computation proper. Computationalism broadly need not even be limited to just digital computationalism as is shown in Chapter 8. As well, Piccinini argues that the difference between computing systems that execute programs and those that do not is important not only to computer science, but also to CTM. He asserts that the mechanistic account allows systems to be described as rule following. Accordingly, it may support CTM that appeal to cognitive explanations in terms of following rules. Similarly, the instructional information processing account proposed in Chapter 6 also does not equate digital computation with the execution of programs. So, at least prima facie, it has a similar advantage for CTM.

Further, if natural cognitive systems do essentially exploit non-discrete magnitudes or quantum effects, etc., then they are essentially *non-digital systems*. Accordingly, they cannot be properly described solely in terms of functions realised by digital computing systems (Scheutz, 1999, p. 192). We should not reject possibly adequate accounts of concrete digital computation for not doing justice to CTM. Instead, if the best account of digital computation is not compatible with CTM, then it is the latter that has to be either revised or rejected.

A weaker version of Smith's conceptual criterion is certainly acceptable for determining the adequacy of candidate accounts of digital computation. To the extent that this criterion does not mandate the provision of a new theory of ontology, an adequate account of concrete computation should be conceptually rich for explaining key notions, such as 'interpretation', 'compilation', 'program', 'algorithm', 'data' and 'architecture'. In short, the conceptual analysis should be grounded in theoretical computer science and computational practice, rather than in metaphysics. It should refer to notions that are commonly invoked by those disciplines devoted to the analysis of computability and the design and construction of real-world computing systems.

As much as is practical, an adequate account should do justice to existing real-world computing systems. This methodological principle is reflected by the dichotomy criterion. Furthermore, real-world computing systems are constrained, on the one hand, by computability and complexity limits, defined in terms of classical computability theory. On the other hand, they are constrained by the laws of classical physics. These constraints motivate the implementation criterion.

In sum, the following criteria should be used for judging the adequacy of accounts of concrete digital computation.

1. The conceptual criterion – an adequate account has to explain fundamental concepts at the heart of computation, such as 'interpretation', 'compilation', 'program', 'algorithm', 'data' and 'architecture'.

2. The implementation criterion – an adequate account has to explain how abstract computations are implemented by physical systems.

3. The dichotomy criterion – an adequate account has to distinguish paradigmatic computing systems from non-computing systems.

4. The miscomputation criterion – an adequate account not only has to cater for successful computations, but also for the possibility of miscomputation.

5. The taxonomy criterion – an adequate account has to explain why different classes of digital computing systems have different computational powers.

6. The program execution criterion – an adequate account has to explain how concrete digital computation and program execution are related.

Finally, whilst Smith concludes that there will never be an adequate account of computation, we shall not reject the possibility of such an account. There is, arguably, no compelling reason to renounce every attempt to provide an account of concrete digital computation. The recommended six criteria are sufficiently constraining to ensure that not almost every account would be deemed acceptable. At the same time, they are not over-demanding so as to preclude any candidate account. In the following chapters these criteria are put to the test to determine which accounts, if any, are adequate for explaining concrete digital computation. In some cases, when an account fails to meet a particular criterion we can try different avenues to correct it, but in others, it may be cheaper to start afresh.

References

Bishop, J.M.: A Cognitive Computation Fallacy? Cognition, Computations and Panpsychism. Cognitive Computation 1(3), 221–233 (2009), doi:10.1007/s12559-009-9019-6

Chalmers, D.J.: Does a rock implement every finite-state automaton? Synthese 108(3), 309–333 (1996), doi:10.1007/BF00413692

Chomsky, N.: Syntactic structures, 2nd edn. Mouton de Gruyter, Berlin (2002)

Chrisley, R.L.: Why everything doesn't realize every computation. Minds and Machines 4(4), 403–420 (1994), doi:10.1007/BF00974167

Copeland, B.J.: What is computation? Synthese 108(3), 335–359 (1996), doi:10.1007/BF00413693

Copeland, B.J.: Computation. In: Floridi, L. (ed.) The Blackwell Guide to the Philosophy of Computing and Information, pp. 3–17. Blackwell, Malden (2004)

Fresco, N., Primiero, G.: Miscomputation. Philosophy & Technology (2013), doi:10.1007/s13347-013-0112-0

Hempel, C.G.: Aspects of scientific explanation: and other essays in the philosophy of science. Free Press (1965)

Piccinini, G.: Computing Mechanisms. Philosophy of Science 74(4), 501–526 (2007), doi:10.1086/522851

Piccinini, G.: Some neural networks compute, others don't. Neural Networks 21(2-3), 311–321 (2008a), doi:10.1016/j.neunet.2007.12.010

Piccinini, G.: Computers. Pacific Philosophical Quarterly 89(1), 32–73 (2008b), doi:10.1111/j.1468-0114.2008.00309.x

Popper, K.R.: The Logic of Scientific Discovery (14th Printing). Routledge, London (2002a),
http://public.eblib.com/EBLPublic/PublicView.do?ptiID=254228

Popper, K.R.: The poverty of historicism. Routledge, London (2002b)

Putnam, H.: Representation and reality. The MIT Press, Cambridge (1988)

Scheutz, M.: When Physical Systems Realize Functions... Minds and Machines 9(2), 161–196 (1999), doi:10.1023/A:1008364332419

Shagrir, O.: Content, Computation and Externalism. Mind 110(438), 369–400 (2001), doi:10.1093/mind/110.438.369

Sloman, A.: What cognitive scientists need to know about virtual machines. In: Taatgen, N., van Rijn, H. (eds.) Proceedings of the 31st Annual Conference of the Cognitive Science Society, pp. 1210–1215 (2009)

Smith, B.C.: Age of significance: State of the art (forthcoming),
http://www.ageofsignificance.org (retrieved)

Smith, B.C.: On the origin of objects. MIT Press, Cambridge (1996)

Smith, B.C.: The foundations of computing. In: Scheutz, M. (ed.) Computationalism: New Directions, pp. 23–58. MIT Press, Cambridge (2002)

Smith, B.C.: Age of significance: Introduction (2010),
http://www.ageofsignificance.org (retrieved)

Ulmann, B.: Analog Computing. Oldenbourg Verlag (2013)

Von Neumann, J.: The general and logical theory of automata. In: Taub, A.H. (ed.) Collected Works, vol. V, pp. 288–328. Pergamon Press, Oxford (1961)

Winograd, T., Flores, F.: Understanding computers and cognition: a new foundation for design. Ablex, Norwood, N.J. (1986)

Chapter 3
Starting at the Beginning: Turing's Account Examined

3.1 Turing Machines from Mathematical and Physical Perspectives

Computability as an abstract notion, underlying any physical digital computation, was the focus of intense interest during the 1930s by mathematicians such as Gödel (1931), Turing (1936), Church (1936) and Post (1936). According to Turing's analysis, digital computation is the process of effectively calculating the values of a function by some purely mechanical procedure executed by a TM. He argued that computation is reducible to carrying out a calculation by following instructions.

Turing's model of computation was inspired by human computation, which could be completely mechanised by breaking its steps into a series of basic operations. The TM was based on a human computor that may facilitate her calculations by using a pencil and a notebook. By following a given set of instructions, she may alter a specific page and turn to another one. The alphabet of symbols available to her can be assumed to be finite. Without loss of generality, the content of each page can be considered as being encoded by a single symbol.[1] By envisaging a large enough notebook with its pages being placed side-by-side, thereby forming a single running paper tape, the tape may be arbitrarily long. The tape may be viewed as consisting of squares that can either be blank or bear a symbol. At each stage of her calculation, the human computor can alter the scanned square, turn her attention to the next, or previous, square or look at the instructions to see which one has to be followed next. At any given moment, she can focus her attention only on a single square.

Turing's analysis of computability was carried out primarily from a *mathematical* standpoint, but also from a *mechanical* standpoint.[2] It studied what

---

[1] Even if the content of each page had to be encoded by more than one symbol (but, say, no more than ten), then that would simply result in a slower TM. Nothing else in the TM would change fundamentally.

[2] The focus of this book is on accounts of concrete digital computation, so whenever Turing's analysis is examined as a basis for *physical* computation, it is referred to as Turing's 'account'. 'Analysis' refers to Turing's formalism of *computability*.

N. Fresco, *Physical Computation and Cognitive Science*,
Studies in Applied Philosophy, Epistemology and Rational Ethics 12,
DOI: 10.1007/978-3-642-41375-9_3, © Springer-Verlag Berlin Heidelberg 2014

functions on the natural numbers are computable, but also what can be done by mechanically following a procedure. Although Turing analysed *mathematical* computability, rather than *concrete* computation, his analysis should be examined as the basis for *any possibly adequate* account of concrete computation. Turing achieved a metamathematical proof that the iteration of some basic atomic operations is sufficient for computing any computable function in mathematics – as we know it today (Agassi, unpublished). In that respect, Post's machines, Kleene's formal systems, Gödel's recursive functions and Church's lambda calculus also extensionally identify the *same class of functions* as computable.

These formalisms are subdivided (amongst other things) in terms of algorithmic versus non-algorithmic models. Gödel's recursive functions, for example, are essentially non-algorithmic. From a strictly intensional viewpoint, recursive functions and recursion theory do not analyse anything about what *computers* can or cannot accomplish at all (Soare, 1996, p. 307). On the other hand, TMs, Church's lambda calculus and Post's machines have an algorithmic interpretation inherent to the computation of a function $F(x)$.

A TM has the following characteristics (Turing, 1936, pp. 231–232).

1. The machine has a read/write head moving along the tape capable of only a finite number of "conditions" (q_1, q_2,..., q_n). These conditions are called "m-configurations" and constitute the machine's states.
2. The machine is supplied with an unlimited running tape divided into squares.
3. At any given time the machine's head scans only one square, which may bear a symbol (referred to as the scanned square and the scanned symbol, respectively).
4. The machine can write a new symbol after erasing any existing scanned symbol.
5. The machine can shift its head one place to the right or left of the last scanned square.
6. The machine can change its state following each atomic operation (e.g., move head one square to the right, scan a square or write a symbol).
7. The machine's behaviour is completely determined by an ordered pair consisting of the machine's state and the scanned symbol.

Let us pause briefly to consider a striking resemblance of the TM to a Post worker (to which we refer as a Post machine). The Post machine for solving computable problems operates in a "symbol space" according to "a fixed unalterable set of directions [i.e., instructions], which [...] determine the order in which those directions are to be applied" (Post, 1936, p. 103). The "symbol space [...] consist[s] of a two way infinite sequence of [...] boxes" (Post, 1936, p. 103), each of which can be either marked or empty. The Post machine is capable of performing one of the following primitive operations (Post, 1936, p. 103).

1. Marking the box it is in (assumed empty).
2. Erasing the mark in the box it is in (assumed mark).
3. Moving to the box on its right.
4. Moving to the box on its left.
5. Determining whether the box it is in, is or is not marked.

Thus characterised, it is easy to observe the resemblance between the TM and the Post machine. The latter moves and works in a symbol space that may be infinite, and the former moves along a possibly infinite tape. Both machines can only be in and operate in a single box/square at a given time (see the third characteristic of the TM above and (Post, 1936, p. 103)). This is the finiteness condition on the number of symbols "observed" at any given moment. The Post machine can only perform any of a finite set of primitive operations (enumerated 1 through 5 above) and likewise the TM (see the TM's characteristics enumerated 3 through 5 above). The TM starts from some initial m-configuration (Turing, 1936, p. 232) and the Post machine starts from a specific box that is "singled out and called the starting point" (Post, 1936, p. 103). Both Turing and Post stipulated that their machines operate according to a finite, numbered sequence of instructions. Although Post did not offer a proof in his 1936 paper, "Finite Combinatory Processes - Formulation 1", he conjectured that his formulation would "turn out to be logically equivalent to recursiveness in the sense of the Gödel-Church development" (1936, p. 105), and, clearly, to Turing computability.

Given that our present focus is on concrete computation, it should be emphasised that both the TM and the Post machine are abstract mathematical objects. A TM can be formally defined as a 7-tuple (Hopcroft et al., 2001, p. 319). TM = $(Q, \Sigma, \Gamma, \delta, q_0, B, F)$ where:

- Q stands for the finite set of states of the TM.
- Σ stands for the finite set of input symbols.
- Γ stands for the complete set of tape symbols and is a superset of Σ.
- δ stands for the transition function, which can be described as $\delta(q, X)$ where $q \in Q$ and $X \in \Gamma$. If the value of $\delta(q, X)$ is defined, then it is a triple (p, Y, D) where:
  - $p \in Q$ is the next state.
  - $Y \in \Gamma$ is the new symbol written in the scanned square (replacing whatever symbol was previously there).
  - D is the direction (either left of right) in which the TM's head should move next.
- $q_0 \in Q$ stands for the initial state.
- B stands for a blank symbol.
- F stands for the set of final states.

To explain further how a TM works, let us consider a 5-state TM for computing the successor function $F(x)$ mod 256, for $0 \leq x < 255$. For simplicity, let us restrict the TM's tape to just ten squares with only three possible symbols: '0', '1' and 'B' (for *blank*).[3] We impose four additional conventions on the TM regarding its configuration. First, the input numeral inscribed on the TM's tape (as well as the resulting output numeral) is delimited by the 'B' symbol on both ends. Second, the

---

[3] The operation of this TM may be simplified, for instance, by using a unary representation of numbers and thus not requiring the '0' symbol. But then, whilst less symbol-types and transition rules are needed for its operation, there is a need for some extra tape to compute the same number.

input numeral is inscribed on the tape in the conventional binary notation. Accordingly, the byte-representation of zero, for instance, would be inscribed on the tape as the numeral 'B00000000B' and the number nine would be represented as 'B00001001B'. Third, when initialised, the TM's head is positioned on the leftmost occurrence of either a '0' or a '1' (and also when it halts).[4] Four, only one binary numeral is inscribed on the TM's tape at any given time.

So, when started in the standard configuration on the tape representing a number x in binary notation the TM halts in the standard configuration representing the number $x+1$. State s_0 is used to scan to the first occurrence of a 'B' to the right of a single block of '0's and/or '1's. State s_1 is used to overwrite the rightmost occurrence of either a '0' or a '1' by either a '1' or '0', respectively. State s_2 is used to overwrite any occurrences of '1's in a single block by a '0'. Once the TM enters the state s_3 it simply scans the first occurrence of a 'B' to the left of any '0's and/or '1's. Once the 'B' is found, the TM's head moves one square to the right and halts in state s_4. Table 3.1 illustrates the operation of this TM by describing its instructions table.

Table 3.1 This instructions table describes a TM that computes the successor function mod 256, for $0 \leq x < 255$. 'E' stands for erase the scanned symbol, 'R' stands for move one square to the right and 'L' stands for move one square to the left. The symbols in the "Operation" column represent operations that are performed successively.

m-configuration	Scanned symbol	Operation	Final m-configuration
S_0	0	R	S_0
S_0	1	R	S_0
S_0	B	L	S_1
S_1	0	E, 1, L	S_3
S_1	1	E, 0, L	S_2
S_2	0	E, 1, L	S_3
S_2	1	E, 0, L	S_2
S_3	0	L	S_3
S_3	1	L	S_3
S_3	B	R	S_4

The example above makes it easy to see that from a purely mathematical perspective, a TM is an abstract transition function. It is figuratively described as a machine that reads and writes symbols in the process of computing that function. But it is "no more a *machine*... than a model aeroplane is an aeroplane, although like any model, its theoretical utility lies in the fact that it has certain properties

---

[4] The usual convention is to have the head positioned over the leftmost 'B' at the start. For simplicity, we change this convention slightly.

that stand in a *systematic relationship* to the things it models" (Seligman, 2002, p. 210, italics added).

Although a TM is an idealisation of a machine, it may be viewed as a simple model of a digital computer.[5] As a model, it was sufficiently simple to facilitate building actual instances according to this design (Israel, 2002, p. 193). But Turing's analysis does not take into account important physical constraint considerations. Firstly, the TM does not utilise energy and no energy dissipates from it, though moving its head to the left or right, reading and writing a symbol on a square and so on – are all *physical actions* with causal effects. For example, it is well known in information theory that data erasure in physical systems necessarily involves the production of heat and, therefore, an increase in thermodynamic entropy (Bais & Farmer, 2008, pp. 633–634).[6] Secondly, unlike physical computing systems, it never breaks down, as remarked in the previous chapter. Thirdly, it has an infinite tape at its disposal, whereas a physical computing system does not have anything that corresponds such infinite tape. The five requirements for a system to compute (listed in Section 3.2) should be viewed as regulative principles and be supplemented to ensure that physical computing systems obey the laws of physics. Notably, the Gandy-Sieg account, which is examined in Chapter 7, considers the implications of the laws of physics for *physical* computing systems.

Perhaps his greatest contribution to the development of digital computing systems was Turing's concept of a universal machine. Turing showed that

> "a single machine [...] can be used to compute any computable sequence. If this [universal] machine [...] is supplied with a tape on the beginning of which is written the [standard description] of some computing machine M [...] then [it] will compute the same sequence as M" (1936, pp. 241–242).

This UTM gives rise to the concept of universality, which unifies data and programs, by using a single TM that takes the program of some other simulated TM M plus M's input as data. Programs and data in the machine's memory may be treated interchangeably. Hence, the single UTM may be viewed as a stored-program digital computer. A UTM U is easier to conceive as a multi-tape TM, rather than the standard single-tape TM. The transition rules of M are stored on the first tape along with the input to M. A second tape of U holds the simulated tape of M (using the same format as for the code of M). A third tape of U is used to hold the current state of M. The last tape of U is used as a scratchpad in case extra space is needed for new values by copying some portions of another tape onto the scratch tape.

The operation of U proceeds as follows. Firstly, the input to M is examined to ensure that it is legitimate for any TM. If it is illegitimate, then U halts without accepting. Secondly, the second tape of U is initialised to contain the input to M in

---

[5] Admittedly, the later Register Machine formalism is a more appropriate model of the modern digital computer, which manipulates data and instructions stored in registers, than the TM, which has to scan the data back and forth along its tape (Soare, 1996, p. 298).

[6] We discuss this principle in more detail in Chapter 6.

its encoded form. Thirdly, 0, as the initial state of M, is inscribed on U's third tape and the head of U's second tape is moved to the first simulated cell of M. Only then may U begin simulating the moves of M by searching on U's first tape for a transition that M would have made for the simulated state and the tape symbol. If no such transition is found, then U halts (as M would have halted). For all cases that M would have entered its accepting state, so does U.[7]

Importantly, the UTM introduces another way for arranging a TM to act in accordance with a program. The head of a special-purpose TM may be "programmed" by means of "internal wiring". Before the special-purpose TM may be used to compute a function, its head must first be "programmed" by "hardwiring" its instructions table. The input has to be then inscribed on the TM's tape. Next, the head needs to be placed over the first input symbol (i.e., the leftmost input symbol) and only then may the TM be set in motion.

In the case of the UTM, on the other hand, the instructions of the special-purpose TM are first translated into, say, a binary code. This code is then inscribed on the UTM's tape, or one of its tapes as in the description above. Turing showed that there exists a universal program UP such that if the head of the UTM is programmed in accordance with UP and provided that any TM M is inscribed on the tape of the UTM, then the UTM will behave as though its head had been programmed in accordance with M.[8] The purpose of the computing system is, thus, modified by storing a program of symbolic instructions in its working memory, rather than by changing its "internal wiring" (Copeland, 1997, pp. 691–692).

The difference between a special-purpose TM and a UTM in the context of program execution requires some further explication. A special-purpose TM corresponds to a conventional computer program. The initial symbols inscribed on the TM's tape as well as the symbols it prints out in the computation of a number correspond to the program's input and output, respectively. Accordingly, from a certain perspective, a UTM may be viewed as a general-purpose computing system that executes programs.[9] However, it is less obvious to view a special-purpose TM as *executing* a program. Instead, we might simply describe it as *acting in accordance* with a program, that is, its table of instructions.

This choice of terminology also leads to a further distinction between a *stored-program* computing system and a *program-controlled* computing system. A UTM is a soft-programmable system, which can simply be reprogrammed by erasing its existing program (TM_1) from its tape and inscribing another one (of some $TM_2 \neq TM_1$) on its tape. Alternatively, we can conceive a multi-head multi-tape UTM that is capable of simulating multiple TMs without first erasing those

---

[7] For a complete description the interested reader is referred to Hopcroft et al. (2001, pp. 377–379).

[8] There are in fact arbitrarily many different programs that can do the work of UP and, hence, many distinct UTMs (albeit all equivalent in their computational power) (Copeland, 1997, p. 692).

[9] Clearly, a UTM can also be viewed in itself as a program, which simulates (or interprets) other programs.

programs (similarly to conventional digital computers).[10] At any rate, the UTM is an idealised model of a stored-program computer[11]. The special-purpose TM, on the other hand, may be only "hardwired", that is, hard-programmed, for a specific computational task by modifying the machine head's internal wiring (figuratively, by means of a plug-board arrangement). Hence, the special-purpose TM may be viewed as an idealised model of a *program-controlled*, not a stored-program, computing system.

Despite the fact that a UTM can, in principle, simulate any conventional digital computer (Hopcroft et al., 2001, pp. 356–361), conventional TMs are arguably inadequate to explain *online* computation. As noted in Chapter 1, an oracle TM may be a better model of online computation. But first, let us briefly describe what an oracle TM is. Oracle TMs were introduced by Turing in 1939: "[t]hese machines may be described by tables of the same kind as those used for the description of [automatic] machines, there being no entries, however, for the internal configuration [that is unique to the oracle]" (1939, p. 173). An oracle TM consists of a standard TM supplemented with a primitive operation, or several primitive operations, that returns the value(s) of an uncomputable function on the integers. This primitive operation is made available by some additional component of the TM that Turing dubbed an "oracle" (Copeland, 1997, p. 705), which works using some "*unspecified* means of solving number-theoretic problems" and "*cannot* be a machine" (Turing, 1939, pp. 172–173, italics added).

Moreover, the oracle is invoked in a *special* way to give the value of the uncomputable function. Calls to the oracle are made by means of a unique state – the "call state", as well as a unique symbol – the "marker symbol". Occurrences of the marker symbol on the tape indicate the beginning and the end of the input string to the oracle. When an instruction in the machine's table puts the TM into the call state, the input is delivered to the oracle. The oracle returns the value of an uncomputable function, such as the halting function H(x, y). The call to the oracle ends with the oracle putting the TM in either the 0-state or the 1-state according to whether the value of the function is 0 or 1 (Copeland, 1997, p. 705).

Given this brief exposition we now return to the claim that an oracle TM is a better model for online computation. Oracle machines, as well as standard TMs, were intended to delineate classes of mathematical problems, to which the issue of the possible *physical* implementation of the machine's logical components is irrelevant (Copeland, 1997, p. 706). Turing acknowledged that oracle machines cannot be constructed (1939, p. 173). But when we extrapolate from Turing's

---

[10] Yet, this construction introduces the problem of synchronicity, which was later tackled by Gandy.

[11] Piccinini argues that, strictly, UTMs should not be viewed as stored-program computers, contrary to the received view (2008, pp. 55–56). For, unlike the stored-program computer, the UTM's tape is used as both an input/output device *and* a memory. Also, he adds that Turing originally considered the machine's internal state, rather than its tape, a memory component (this capacity is discussed in Section 3.2 in relation to the second requirement). If that is right, then the UTM's internal states are short-term memory, whereas a conventional stored-program computer requires a long-term working memory.

analysis regulative principles for the guidance of an account of concrete computation, questions of physical realisation are paramount. To reprise, an online computational process is one that interacts with its environment. Conventional TMs lack this online capacity, for they begin with a fixed program and a fixed input and proceed without further external input until they halt, if they ever do. A TM equipped with an oracle component, in contrast, typically, has a fixed program at its core, and the oracle component is the mechanism for communicating with the environment (e.g., a database or the World Wide Web) (Soare, 2009, p. 387).

Some argue that the oracle TM is a formalisation of another type of TMs introduced in (Turing, 1936): the choice TM (Eberbach et al., 2004, p. 164). Turing only briefly mentioned choice TMs as "machines [...] whose motion is only partially determined by [its] configuration" (1936, p. 232). An important similarity between choice TMs and oracle TMs is their interactive mode of computation: they both make queries to some external agent during the course of the computation. Turing's oracles were meant to represent uncomputable information sourced from outside the TM. In the case of the choice TM, the external agent is some user, such as a human user of a computer. In that sense, the oracle TM and the choice TM represent an alternative model of computation to the one offered by a conventional TM, which "operate[s] in a closed-box fashion as if on 'automatic pilot'" (Eberbach et al., 2004, p. 163). When a choice TM "reaches one of [possible] ambiguous configurations, it cannot go on until some arbitrary choice has been made by an external operator"(Turing, 1936, p. 232).

In what sense then does a choice TM or an oracle TM represent a model of online computation? Whilst modern digital computers do not typically invoke any non-algorithmic "oracle" processes for "guessing" the right answers to some uncomputable problems, the oracle TM has some import for online computing systems. It introduces a mechanism by which a digital computing system interacts with its environment, thereby surpassing the standard model of offline computation. Nevertheless, both the choice TM and the oracle TM are a limited model of online computation. An important feature of online computation is that of input coming at arbitrary times and intervals. Consider a router in a network that calculates the destinations of messages to be sent by using its routing table. Some router links may go down and then come back up at arbitrary times. The program for computing the routing table, therefore, has to be "online". Neither the choice TM nor the oracle TM has an analogous feature. They only "consult" the external agent when they reach a certain configuration.

With that in mind, we turn next to distill the key requirements implied by Turing's account for physical systems to compute. For our purposes, neither the oracle TM nor the choice TM model fundamentally changes the overall picture in terms of Turing's analysis as a precursor to any account of concrete digital computation. Yet, to extend his analysis to also apply to real-world online computing systems, one more specific requirement is included below resulting from the oracle/choice TM model.

3.2   The Key Requirements According to Turing's Account

According to Turing's account, digital computation consists of sequences of operations on symbols that can be performed by either human computors or by some mechanical devices following a *finite* number of fixed rules. The key requirements for a physical system to perform digital computation, by Turing's lights, are as follows.

- 1st Requirement: having the capacity to recognise symbols.
- 2nd Requirement: having the capacity to write symbols and store them in memory.
- 3rd Requirement: having the capacity to change states.
- 4th Requirement: having the capacity to follow instructions (Turing, 1936, pp. 231–232).
- 5th Requirement: having the capacity to interact with its environment during processing (Turing, 1939, pp. 172–173).

The first fundamental requirement of Turing's account is the system having the capacity to recognise symbols. Any computation executed by the system cannot proceed without the system identifying the scanned symbol and responding accordingly, for example, by erasing a symbol or writing a new one. The TM's head moves back and forth along the tape and scans *finitely* many basic symbols. When considering a physical instantiation of a TM, using *infinitely many* symbols can result in some of those symbols being too similar to the others to be reliably distinguished from them in finite time (Parikh, 1998). When considered as a purely mathematical object, a TM that uses infinitely many symbols introduces a mathematical problem. How can the entire instructions table be specified if there are infinitely many symbols? This requires an infinite program.

Importantly, nothing essential hinges on TM's symbols *being* symbols *per se*. The only relevant property of these symbols is that they be recognisable and modifiable by the computing system (Seligman, 2002, pp. 217–218). Instead, we may replace 'symbols' with 'marks', where the latter simply means a "single, repeatable and readily *identifiable* [figure], such as a letter [...] or a circle, a square, a star, etc." (Kearns, 1997, p. 273, italics added). The marks need not even be discriminated visually. Consider, for example, the Braille writing system in which marks may be discriminated non-visually. The marks simply need to be sufficiently distinct to allow their discrimination from a finite set of possible marks. In Chapter 6, it shall become apparent that these symbols may simply be identified with *data*. The key point is that there is no necessity for any *meaning* to be attached to these marks. If there is any, it is purely fortuitous (Kearns, 1997, p. 281). Accordingly, a TM may be described as manipulating marks (data) for producing a sequence of marks (data) from a single finite set of possible marks (data) (Kearns, 1997, p. 274).

Smith pushes this point further and dubs Turing's analysis "a theory of marks" (2010, p. 27). On his view, Turing's analysis is akin to a mathematical theory of differential causality. Namely, it is a theory about what configurations of marks, or concrete *stuff*, can be moved around and/or transformed by finite physical processes into some other configurations of marks. Smith concludes that Turing's analysis is *not* a theory of *computation*, since it is not a theory that applies *only* to digital computation *qua* digital computation (2010, p. 28).

Be that as it may, a TM unable to recognise, or identify, symbols/marks will not be able to read symbols/marks off the tape rendering it incapable of successfully completing its operation.[12] The first requirement may be broken down further into four conditions. Raymond Nelson lists four conditions that an *FSA* has to satisfy *to count as accepting* a symbol (1982, pp. 166–170).[13] The first three are essential for symbol recognition in idealised TMs. The first condition, *universality and discrimination*, is that the FSA must be able to assign types to tokens (or universals to particulars) over a potentially infinite domain. If an FSA can accept a particular symbol, it must be able to discriminate it among many types over a very large, though finite, domain. Similarly, a *TM* must be capable of discriminating all the types of symbols that may be inscribed on its tape.

The remaining conditions are the recognition of *One-Token-Many-Types*, the recognition of *One-Type-Many-Tokens* and lastly the recognition of *degraded tokens*. The second condition implies that the acceptance relation be merely relational rather than functional, as the FSA has to discriminate among different patterns made up of the same parts. It should be able to assign more than one distinct type to a single set of tokens. This condition requires that the acceptance of a set of tokens of more than one type is computationally possible (Nelson, 1987, p. 603). The third condition states that the FSA must be able to accept the tokens of a single type from any one of several disjoint sets of tokens. The FSA must be capable of accepting individuals having deviant properties as being of the standard/acceptable type (Nelson, 1987, p. 601). Unlike the first condition, the third one requires that there is a family of disjoint sets and one and the same type is the type for every set of that family. The fourth condition requires a token to be recognised, even if its individuating properties are blurred (e.g., a degraded ß as an instance of the symbol type B), by some means of correction based on the context. Whilst TMs are idealised and, hence, not subject to any possible degradation of symbols, this is an important consideration in *physical* systems that are subject to physical *noise*.

The second key requirement of Turing's account is the system having the capacity to write symbols and store them in memory. A TM reads symbols, but it also writes symbols for intermediate calculations and for printing the output. The machine's tape is a general-purpose storage, which serves both as a long-term working memory and the input/output device.[14] TMs must have input for their normal operation, but they do not necessarily produce an output. In the latter case

---

[12] Hereinafter, we shall continue using the term 'symbol' for consistency and to be faithful to Turing's original formulation.

[13] Incidentally, Nelson dubs these four conditions "gestalt conditions" that are supposedly satisfied by most humans (1987, p. 602).

[14] The tape of a TM is (at least in principle) infinite. But equipping TMs with infinite memories was simply intended to show that some mathematical problems cannot be solved by such machines even given unlimited running time or memory space (Copeland, 1997, p. 691).

they accept the input, perform some operation and halt. To emphasise, a TM need not have a single tape. It can have multiple tapes with a finite sequence of input symbols on one tape. Yet, single-tape TMs are equivalent to multi-tape TMs in terms of their computational power: any multi-tape TM can be simulated by a single-tape TM (Hopcroft et al., 2001, pp. 336–338). (It is worth noting that there is no requirement that the TM have the capacity to *change* symbols, that is, other than changing from a blank to some symbol.)

Moreover, without the capacity to store symbols in memory, TMs reduce to FSA with bidirectional input tapes. For a TM is an FSA with a single infinite tape for reading and *writing* symbols (Hopcroft et al., 2001, p. 317), whereas an FSA only has a unidirectional input tape and no output tape.[15] In the above 7-tuple notation of a TM, only Γ and B are elements that are added to the 5-tuple notation of the FSA and the transition function is of the type: $Q \times \Gamma \rightarrow Q \times \Gamma \times \{L, R\}$; L, R stand for moving one cell left or right on the tape (Israel, 2002, p. 193). FSA clearly compute, yet, their computational power is limited compared with that of TMs, due to the lack of a writable memory medium in an FSA (Wells, 1998, p. 287).

The third key requirement of Turing's account is the system having the capacity to change states. A TM can change its state or remain in its current state in accordance with three *neighbourhood* conditions. First, the TM can change the symbol only in a scanned square and then at most *one* symbol. Second, the TM can move to a different set of scanned squares, but only within a certain *bounded distance* of the scanned square. Third, any atomic operation must depend only on the TM's *current state* and the *scanned symbol* (Soare, 1996, p. 292). The TM can be in any of a finite number of states (formally denoted above by Q). As a function of its current state and the symbol just scanned, it can either change its state or remain in the same state it was in. Again, this requirement applies equally well to an FSA, which either changes its state or not depending on the symbol read.

The fourth key requirement of Turing's account is the system having the capacity to follow instructions. This is a fundamental requirement of any computing system whatsoever, by Turing's lights. This point was abundantly clear in Wittgenstein's critique of Turing's analysis of computability (Shanker, 1998, pp. 9–10). The answer to the question 'how does a TM go about computing X?' is constituted by Turing's account of the mechanics of the system, which is governed by following an instruction. "Calculation depends on our brains following a set of simple mechanical rules, which are such that they can also be followed by a [Turing] machine" (Shanker, 1998, p. 10). The computation of the TM depends on its ability to follow the sub-rules of its program, that is, the TM's table of instructions.

To reprise, the ability to follow instructions is exercised in a slightly different manner by special-purpose TMs and UTMs. Turing did not explicate the mechanism that causes the machine to follow these instructions. But in the sense that TMs are merely idealisations, it was not essential that he explicate such a

---

[15] Strictly, a TM also has a head for scanning symbols on the tape and a state register, whereas an FSA is devoid of any such structure.

specific mechanism, which causes the machine to follow instructions (Copeland, 2004, pp. 5–6). Whereas special-purpose TMs can only be accurately construed as *acting in accordance* with instructions, which are hardwired – or implicit, UTMs also *follow* instructions *explicitly*. Accordingly, a UTM can carry out any computation that can be carried out by any special-purpose TM, thereby yielding a powerful kind of *programmable* computing systems.

Finally, the fifth key requirement of Turing's account, which is not derived from the conventional TM, is the system having the capacity to interact with the environment during processing (Turing, 1939, pp. 172–173).[16] This requirement is derived from oracle-type TMs. It is *only applicable* when we analyse *interactive* or *online* computing systems. Again, the crucial point is not about the oracle TM finding the values of uncomputable functions, but about it being a model of computational processes that do not run uninterrupted once the entire input was provided.[17] To accomplish that, the computing system has to either be capable of actively querying the environment for more input or be somehow "interrupted" by the environment when a new input becomes available. Strictly, the oracle TM is only compatible with the former mode of interaction, for once the TM enters the unique call state it is the TM that presents a question to the "oracle".

Possible examples for such online computing systems are automatic teller machines (ATMs), parallel distributive systems and computer games. ATMs typically run various software modules for security authentication, user identification, currency conversion, query and update user accounts and so on. To allow multiple users to use the ATM it runs in a continuous mode awaiting some new requests. Each such request is an interrupt from the environment to query some user's account, withdraw cash, deposit a cheque, etc. Similarly, in parallel distributive systems consisting of multiple computing systems, a system CS_1 may interrupt another system CS_2 either requesting some data from CS_2 or providing it with some new data that were processed/generated by CS_1. In like manner, computer games are interactive, by definition, in that they consist of positions and moves played by a player and an opponent (or a system and its environment). Computational processes in games proceed in a question-and-answer mode where the entire input cannot be simply provided at the start of the game.

3.3 Turing's Account Evaluated

An important disclaimer should be offered first. Turing's analysis is not and was never intended as an account of *concrete* digital computation. Nevertheless, whilst TMs are *idealised* models of digital computation, to a great extent they provide the

---

[16] Incidentally, an invocation of the oracle black box by the TM is a single atomic operation. Still, this characteristic does not present a serious limitation to the oracle TM adequately modelling online computation.

[17] In being capable of solving problems that are uncomputable, the oracle TM is infinitely more powerful than any conventional digital computing system. To avoid any doubt, the intention here is not to consider the oracle black box as anything resembling a basic component of a digital computer.

ground rules for both computer science and computational practice. An analysis of computational problems in terms of TMs can give guidance to programmers about what they can or cannot be able to solve by programming a computer (Hopcroft et al., 2001, p. 316). Importantly, UTMs may also be viewed, to a considerable degree, as a blueprint for all conventional real-world digital computers. Since a TM is the quintessential computing system, it is well-worth exploring in the context of concrete computation as well.

That said, we now turn to examine the reasons behind the claim that Turing's account is inadequate as an account of *concrete* computation. One incontrovertible reason for this claim is that this account fails to meet the implementation criterion. This reason is incontrovertible, because it is *trivially* true when dealing with the TM as a purely mathematical model of computation. But let us examine this account more broadly by focusing on the key requirements listed above for determining whether a given physical system is computational. The TM was the conceptual basis for the design of von Neumann machines and in principle there is no technical difficulty in constructing a physical TM. The main problem, of course, would be equipping it with an infinite tape (Kearns, 1997, p. 275). However, let us even stipulate that a physical TM is equipped with an arbitrarily large memory as required. Even with this stipulation the requirements above do not immediately lend themselves to the *manner* in which the system has to be physically implemented.

For example, a TM has the capacities to recognise and write symbols. The TM's capacity to recognise symbols may be reduced to the first three conditions proposed by Nelson as discussed above. However, these conditions apply equally well to abstract systems, such as FSA or Abstract State Machines (Gurevich, 2001) and to physical systems, such as flip-flops or the early punched card computers. On punched cards, for example, a 1-bit was represented by a punched hole and a 0-bit was represented by a card position that was not punched. The card reader clearly needed to distinguish the punched hole from the lack thereof. In flip-flops, certain voltage ranges, say, [0v – 0.8v] and [1.7v – 2.5v] represent 0-bit and 1-bit, respectively. The flip-flop has to be capable of distinguishing the two voltage types, within some margins of error, otherwise, it may not yield the correct output.

Turing's account does not specify the necessary causal links between symbol manipulation and physical state changes in the implementing physical system. The third requirement – the system having the capacity to change (computational) states – also says nothing about how they are mapped to the physical states of the implementing system. Yet, in a physically implemented TM, one would expect that physical tokens of symbols would be causally effective in determining just how the TM changes its states and processes inputs to produce the correct outputs (Kearns, 1997, p. 275). The TM has as "one of its defining characteristics the property of being independent from the physical... [and being] defined *without* recourse to the nature of the systems that (potentially) realize [it]" (Scheutz, 1999, p. 163, italics original).

The issue of physical immateriality is also related to the evaluation of Turing's account based on the miscomputation criterion. Although the TM is described as performing physical actions: moving its head, reading, erasing and writing symbols, it utilises no energy. And of course, a TM never breaks down. Turing's account, as is also reflected in the above requirements, does not cater for the possibility of a miscomputation, say, as a result of not being able to use the tape when calculating (approximating) the value of π, for example.

Needless to say, this abstraction from physical constraints is justified when examining a TM from a purely mathematical view, but not so much when extending this mathematical concept to apply to concrete computation. Elsewhere, Turing defines a miscomputation, that is, a computational malfunction, as an error of malfunctioning that is "due to some mechanical or electrical fault which causes the machine to behave otherwise than it was designed to do" (1950, p. 449). He adds that "[b]y definition [abstract machines] are incapable of errors of functioning. In this sense we can truly say that 'machines can never make mistakes'" (Turing, 1950, p. 449).

Since the claim that Turing's account fails to meet the miscomputation criterion seems trivially true, let us make some brief observations on this matter. Nelson's fourth condition, the recognition of degraded tokens, does not apply to idealised TMs. However, that condition is certainly important when dealing with a concrete computing system, the parts of which are unavoidably susceptible to wear and tear. Forces of friction act at the surface of contact of any two moving components of the computing system, such as the rotating discs of computer hard drives or the head of a physically constructed TM that moves and scans squares on the tape. Some of the effects of friction are the production of heat and the wear and tear of physical bodies. So, some data, which were stored in physical memory, may become distorted. The processing of distorted data can induce a miscomputation.

Unlike the noiseless operation of idealised TMs, real-world digital computing systems are always subject to physical noise. Likewise, a physical instantiation of a TM becomes susceptible to noise, which may arbitrarily alter the contents of its tape and or its current state. Accordingly, its actual transition operations can no longer be reliably predicted, even though their probabilities may be calculated. The unreliability of the moving parts used for building digital computers was acknowledged by von Neumann (1961, pp. 303–304). He argued that high reliabilities of and error checking in digital computing systems are crucial. These systems may not only have to perform an arbitrarily large number of steps in a short time, but in a considerable part of their procedure they are permitted not even a single error (Von Neumann, 1961, p. 292). Any single error may proliferate thereby resulting in a miscomputation (cf. the 1994 Pentium FDIV bug[18]).

Turing's account also fails to meet the conceptual criterion. For one thing, nothing in this account sheds light on what an architecture is or what the difference between compilation and interpretation is. As regards an explanation of a computational architecture, Turing's account is indifferent to the manner in which hardware components are interconnected to meet certain functional and

---

[18] This bug was a famous one in Intel's microprocessors. It was caused by an error in a lookup table that was a part of the processor's hardware divide unit. Whilst the bug was considered "rare and data-dependent" (Intel, 2004), many of these defective processors were eventually recalled by Intel.

performance constraints. It certainly requires of any computing system the capacity to write symbols and store them in memory.

However, it says very little about how the architecture of the computing system. In that respect, Turing's analysis is limited to the following two pertinent neighbourhood conditions (Soare, 1996, p. 292) on the system's memory. First, the TM's head can change the symbol only in the scanned square and no more than one symbol. That is, only a particular region of the memory can be acted on at any given time. Second, the TM's head can move to a different set of scanned squares, but only within a certain bounded distance of the scanned square. That is, there is a limitation on the region of memory that can be accessed.

On similar grounds, the difference between compilation and interpretation, for example, cannot be easily explained on Turing's account. Whilst both a compiler and an interpreter are simply programs, they are fundamental in programmable concrete computing systems. A digital computer can only follow instructions if these instructions are specified in some format that is accessible to the computer. The translation of a program into that accessible format, depending on the particular computer architecture, is precisely the objective of compilers and interpreters. Turing's account takes it for granted that the transition rules specified in the TM's table are accessible to the controller to act on. But any constraints that are typically associated with the process of compilation or interpretation of these transition rules are ignored by Turing's account.

Lastly, it is hard to see how Turing's account can meet the dichotomy criterion. For example, most modern digital computers, particularly those based on multiprocessor or multicore architectures, regularly perform multitasking, leading to interaction among processes or threads within processes. This makes it cumbersome to regard the execution of any single program strictly as a conventional *single*-tape, *single*-head TM. For a single program may include multiple threads blurring the classical division of computational labour among sequential procedures.[19] This, of course, is not to say that it is not *possible* to simulate *any* multithreaded program running on a parallel architecture as a single TM. Moreover, stateless trivial computing systems, by definition, do not store data in memory. Surely, we would not classify systems such as AND gates and n-bit adders as non-computational for not being memory based.

3.4 Concluding Remarks: Digitality, Determinism and Systems

Some may argue that faulting Turing's account as an inadequate account of concrete digital computation does him an injustice. For the TM, much like Church's lambda calculus or Post's machines, is a *mathematical* model of computation. That is true

---

[19] It might be argued that all that is needed for Turing's account to apply to multithreaded programs is simply to extend the standard TM model to be multi-tape and/or multi-head. But, strictly, more than that is required. For one thing, the synchronisation mechanism would have to be spelled out. Parallel computation is more adequately analysed by the Gandy-Sieg account examined in Chapter 7.

and, yet, Turing's analysis of computability is foundational to both computer science and computational practice, as we know them today. Besides, the concept of a TM is employed in cognitive science all too often whenever cognition, undoubtedly being a physical phenomenon, is claimed to be computational. It has been the aim of this chapter to clarify the strengths and weaknesses of the TM model when extended to apply to physical systems as well. By doing so, it may also become clearer what the salient traits of a candidate account of concrete computation are (even if it turns out to be inadequate).[20]

Before concluding this chapter, let us pause briefly to remind the reader of the scope of our analysis in the ensuing chapters. We start with the "simpler" term 'system'. Turing's analysis refers to machines that compute. These are conceptual machines, but they can also be physically constructed (*sans* the infinite tape). For Turing, 'effective' meant 'mechanical' and his interest was primarily in the mechanisation of *mathematical procedures*. Most of the interest in the early days of digital computers was in the construction of machines to facilitate brute-force approaches to complex computational problems. Yet, Turing's original analysis was about finding a mechanical method. This method may simply be a book of rules that can be followed by some "mindless" agent. Accordingly, it was essential that TMs required no intelligence to follow their mechanical rules (Shanker, 1998, pp. 15–16). To avoid any overtones of *intelligence* that is often associated with (Turing) *machines*, particularly following his influential paper, "Computing Machinery and Intelligence", a computing 'system' is used throughout this book, rather than a computing 'machine'.

But what is a *system*? A system here is vaguely characterised as a *pseudo-isolated* aggregate of parts that strongly interact within it to accomplish a defined task. This characterisation applies both to physical instances, in which case the interaction is *causal*, such as car engines, printers, screens, aeroplanes and ships, as well as to abstract instances, such as diagrams, conceptual models, economic systems and FSA.[21] The characterisation above suggests that there exists a boundary between the "inside" and "outside" of the system, viz., a boundary that divides the system and its parts from the environment. In the context of concrete computing systems, this is a physical boundary, which is necessary to warrant the label 'system' (Smith, 1996, p. 70).

More specifically, our focus is primarily on *open* systems that interact with their environment. This interaction may be limited in the case of offline computation or ongoing in the case of online computation. As well, it may take the form of "data", "information" (two terms to be qualified in Chapter 6) or energy transfers into or out of the system's boundaries depending on the particular system

---

[20] A notable example is Copeland's algorithm execution account that aims to extend Turing's original analysis so that it does not fall prey to Searle-like or Putnam-like trivialisation in the context of physical systems.

[21] Yet, it is not universally applicable to all systems. The solar system, for instance, can hardly be characterised as the interaction of its parts (i.e., a sun and planets) to accomplish some *task*. The same may be said about microscopic particles that are in some state of interaction without being geared to accomplish some task.

in question.[22] This characterisation is certainly compatible with Putnam's usage of open systems when proving the theorem that every physical object is a realisation of every abstract FSA (Putnam, 1988, pp. 121–125). Still, for our purposes, nothing crucial hinges on this usage of computing *systems*, rather than *machines*. The former simply seems to be more neutral (at least prima facie) on the question of how "intelligent" a system is.

Trying to draw a clear distinction between deterministic and nondeterministic computing systems is a subtler task. Digital computation may be either deterministic or not deterministic (e.g., probabilistic computation or pseudo-random computation). 'Deterministic' in the context of FSA or TMs, for example, simply means that for each input there is one and only one state to which the FSA/TM can transition from its current state (Hopcroft et al., 2001, p. 45). 'Nondeterministic', on the other hand, refers to an FSA or a TM whose transition function takes a state and an input as arguments and returns a set of zero, one or more states (Hopcroft et al., 2001, p. 56). This ability is often expressed as the power to correctly *guess* something about the input. For example, consider a TM for searching for a particular string (e.g., keywords to be matched) in a long text (e.g., an electronic book or a web search). It would clearly be helpful, if the TM were able to "guess" that it were in the beginning of one of those strings and use a sequence of its states to do nothing else but check that the search string appears character by character. Still, any nondeterministic FSA (or TM) may be converted to some deterministic FSA (or TM), though the latter will have many more states (Hopcroft et al., 2001, p. 55).

Most *conventional* digital computing systems are deterministic, to some degree of idealisation, for their behaviour is repeatable and systematic. A dry run of a deterministic algorithm using some test data systematically yields the same output when its input and initial state remain unchanged. Still, to some extent, when subject to physical noise even so-called conventional *deterministic* computing systems are actually *not deterministic*. For there is an element of uncertainty introduced by a possible transient or permanent malfunction of any of the system's components.

The determinism of computer programs is likewise a subtle matter. Possible delays in the process of executing programs may sometimes make its results unpredictable. The multiprocessing (or multithreading) approach in conventional computing systems is susceptible to performance problems, due to unpredictable delays while processes (or threads) are within "critical sections". Some of these unpredictable delays can be the result of caching or paged memory management, which are under the control of the computer's operating system (hereafter, OS) with support from the hardware. These delays can also be the result of the "external" environment, such as input/output devices (e.g., a mouse, keyboard or a USB flash drive) or communication lines to other computing systems (e.g., through modems, old serial ports or Bluetooth enabled devices). Even low-level programming languages (notably, assembler), which have some access to the

---

[22] FSA or idealised TMs clearly do not transfer energy into and out of their boundaries. Still, one may speak of "data" transfers in those cases.

hardware level, cannot always deal with such external noise through the existing error handling mechanisms (White, 2011, pp. 192–195).

Still, for simplicity and following Turing's original analysis, the ensuing discussion mostly focuses on *deterministic* computing systems. Conventional TMs, or *a-machines* in Turing's terminology, are deterministic. Turing stated that "if at each stage the motion of a machine [...] is *completely* determined by the configuration [consisting of its m-configuration and the scanned symbol ...] the machine [is] an 'automatic machine'" (1936, p. 232, italics original). For every possible input and a given initial state, there is only one possible state into which the conventional TM transitions. Every future state transition can, in principle, be accurately predicted by simulating the program being executed without any element of uncertainty or surprise.

Finally, we make some closing remarks on the *digitality* of computing systems. "These are the [... systems,] which move by sudden jumps or clicks from one quite definite state to another" (Turing, 1950, p. 439). It is widely agreed that the states of these systems are *discrete*. Two relational aspects of a digital state seem to be uncontroversial. The first one is that the state has sharp boundaries (Smith, 2005). It implies that there is no room for imprecision or degrees: say, in a binary mode, a system in a state s is *either* idle or active, *either* off or on.[23] What matters in this regard, is the *contrast* among the possible state-types, rather than what they *are*.

Real-world digital systems approximate this principle by spending relatively little time in indeterminate states. For example, in electronic digital systems, which operate on the voltage range between 0v and 2.5v, the *active* state is defined as any voltage greater than, say, 1.7v and the *idle* state is defined as any voltage less than, say, 0.8v. The grey area between 0.8v and 1.7v is indeterminate and when an input signal is in that range, the output value is undetermined. Properly designed, the system should not process any data until it has stabilised into one of its two designated state-types (Tucker, 2004, pp. 16–2).[24]

The second relational aspect of a digital state is that the possible state-*types* are homogenous throughout the whole system (Smith, 2005). If s can either be true or false, then these true and false state-types in a binary system are consistent[25] and interchangeable throughout the system. There are no internal variations in the

---

[23] Haugeland describes this aspect as 'positivity' allowing the "reidentification [of the state to...] be *absolutely* perfect" (1997, p. 10, italics original). The sharp boundaries of the state make it possible to "perfectly" identify the value of the state from some set of possible state-types.

[24] "Perfection is an ideal, never fully attainable [in digital computing systems] in practice. Yet, the definition of 'digital' requires that [whilst] perfection [... is not practicably] possible, [... it is] *reliably* achievable" (Haugeland, 1997, p. 10).

[25] Suppose that some computing system has two types of logic gates. The gates of the first type operate in the conventional mode, where any voltage in the range, say, between 1.7v and 2.5v is interpreted as *1* (or TRUE). But for gates of the second type any voltage in that range is interpreted as *0* (or FALSE). As discussed already in the previous chapter, for these two types of gates to co-exist in a single circuit some buffer has to be added between them. In this manner, the consistency of the possible state-types of these logic gates is preserved throughout the system.

possible types the system may be in when in state *s*. For example, when any pixel on the screen can be either black or white, there is no *partly* black or *almost* black. As a possible state-type, *black* is the same everywhere on the screen. Following from these two aforementioned relational aspects, each digital state of the computing system can be adequately described in finite terms (Gandy, 1980, p. 127).

Nonetheless, the question whether 'digitality' is synonymous with 'discreteness' is controversial. Those who answer it in the negative typically argue that *digitality* also tracks the representation of *numbers* via *digits*. Indeed, "information" in conventional computing systems is traditionally represented, stored and transmitted as groups of bits, that is, in a binary code. This binary code represents both numbers and letters of the alphabets (Bryant & O'Hallaron, 2003, pp. 1–3). Yet, there is no consensus about the representation of numbers being inherent in *every possible* digital computing system.

This disagreement is part of an ongoing debate about *modes of representation*. Nelson Goodman, for one, argued that a digital system has nothing *special* to do with *digits* (or numerals), anymore than an analogue system has something special to do with *analogy* (1968, p. 160). "The dime-counter displaying numerals is a simple example of what is called a digital computer" (ibid). "To be digital, a system", according to Goodman, "must be not *merely discontinuous* but *differentiated* throughout, syntactically and semantically" (1968, p. 161, italics added). By his lights, the characters of a digital system may have objects or events of *any other kind* as their inscriptions. According to this analysis, 'digital' can be treated as synonymous with 'discrete'.

On the other side of the debate, it is argued that digitality requires more than just discreteness (or differentiation). A notable proponent of this approach was David Lewis. He argued that "*digital representation* of numbers [is better defined] as representation of numbers by differentiated multi-digital magnitudes" (Lewis, 1971, p. 327, italics original).[26] On his view, it is insufficient to determine that a representation is *digital* purely on the basis of it being discrete. Instead, digital representation also requires using at least two digits, where both the digit tokens as well as their values are differentiated. Recently, Corey Maley has defended a similar approach characterising digital representation as the following combination (2011, pp. 124–125).

1. Being an ordered *sequence of digits*; and
2. The sequence being in a particular *base* (e.g., binary, octal or decimal) that determines both the value of each digit as a function of its place in the sequence as well as the number of possible numerals, which may be used for each digit.[27]

---

[26] Multi-digital magnitude is defined as "any physical magnitude whose values depend arithmetically on the values of several few-valued unidigital magnitudes" (Lewis, 1971, p. 326).

[27] It is worth noting that Maley's characterisation fails to account for natural numbers in *unary* notation. For the position of any '1' in a sequence of 1's has no special import. So, this characterisation has the unwanted consequence that the above TM for computing the successor function for numbers in unary notation (refer footnote 3) is not considered a digital system.

But though virtually all digital computers are designed using the representation of numbers, this is only by convention and not by necessity. Piccinini, for instance, argues that any account of digitality in terms of modes of representation should be avoided when analysing digital computing systems. For our goal should be to analyse them as *computing* systems rather than as *representational* systems (Piccinini, 2008, p. 48). Piccinini maintains that the notion of digital computation does not inherently require the notion of representation of numbers via digits or anything else (personal communication). Vincent Müller, who seems to share this view, claims that despite there being "cases where the digital states *do* serve [a] representational function [... this] should not be taken as a motivation to re-introduce semantical notions into the definition of digital states or computing" (2008, p. 120, italics original). On his view, a state is digital *iff* it is a token of a type that serves a particular *function*. But digital states are "situated at the level of syntactic description" (ibid) (i.e., being of a certain state-type) and need not involve any number (or other) representation.

Still, whilst *extrinsic* representation is certainly *not* inherent in concrete computation, it is hard to see how *mathematical* representation can be avoided. As remarked in Chapter 1, any system construed as computing some mathematical function represents the arguments and values of that function. Whereas digital computation does not inherently require the notion of *representation of numbers via digits*, it does seem to require the broader notion of mathematical representation. FSA, for example, do not manipulate numbers, but rather strings of characters. Still, any FSA is amenable to a description in terms of a mathematical function that maps some domain onto some range. An OR gate can only be judged as either computing correctly or miscomputing relative to the logical disjunction function it represents. Concrete computation can be viewed as the evaluation of some mathematical function given some input argument(s). The discussion of the representational character of digital computation continues throughout Chapters 5-8.

References

Agassi, J.: The Turing test (unpublished)

Bais, F.A., Farmer, J.D.: The Physics of Information. In: Adriaans, P., van Benthem, J. (eds.) Handbook of the Philosophy of Information, pp. 609–683. Elsevier, Amsterdam (2008)

Bryant, R., O'Hallaron, D.R.: Computer systems: a programmer's perspective beta version. Prentice Hall, Upper Saddle River (2003)

Church, A.: An Unsolvable Problem of Elementary Number Theory. American Journal of Mathematics 58(2), 345 (1936), doi:10.2307/2371045

Copeland, B.J.: The Broad Conception of Computation. American Behavioral Scientist 40(6), 690–716 (1997), doi:10.1177/0002764297040006003

Copeland, B.J.: Computation. In: Floridi, L. (ed.) The Blackwell Guide to the Philosophy of Computing and Information, pp. 3–17. Blackwell, Malden (2004)

Eberbach, E., Goldin, D., Wegner, P.: Turing's Ideas and Models of Computation. In: Teuscher, C. (ed.) Alan Turing: Life and Legacy of a Great Thinker, pp. 159–194. Springer, Heidelberg (2004)

Gandy, R.: Church's Thesis and Principles for Mechanisms. In: The Kleene Symposium, pp. 123–148. North-Holland (1980)

Goedel, K.: Über formal unentscheidbare Sätze der Principia Mathematica und verwandter Systeme I. Monatshefte für Mathematik und Physik 38(1), 173–198 (1931), doi:10.1007/BF01700692

Goodman, N.: Languages of Art: An Approach to a Theory of Symbols. Oxford University Press (1968)

Gurevich, Y.: The Sequential ASM Thesis. In: Paun, G., Rozenberg, G., Salomaa, A. (eds.) Current Trends in Theoretical Computer Science Entering the 21st Century, pp. 363–392. World Scientific, Singapore (2001)

Haugeland, J.: Mind design II: philosophy, psychology, artificial intelligence (rev. and enl. ed.). MIT Press, Cambridge (1997)

Hopcroft, J.E., Motwani, R., Ullman, J.D.: Introduction to automata theory, languages, and computation. Addison-Wesley, Boston (2001)

Intel. FDIV replacement program: statistical analysis of floating point flaw (2004), http://www.intel.com/support/processors/pentium/sb/CS-013005.htm (retrieved January 28, 2012)

Israel, D.: Reflections on Gödel's and Gandy's Reflections on Turing's Thesis. Minds and Machines 12(2), 181–201 (2002), doi:10.1023/A:1015634729532

Kearns, J.T.: Thinking Machines: Some Fundamental Confusions. Minds and Machines 7(2), 269–287 (1997), doi:10.1023/A:1008202117814

Lewis, D.: Analog and digital. Noûs 5(3), 321–327 (1971)

Maley, C.J.: Analog and digital, continuous and discrete. Philosophical Studies 155(1), 117–131 (2011), doi:10.1007/s11098-010-9562-8

Müller, V.C.: Representation in Digital Systems. In: Briggle, A., Waelbers, K., Brey, P. (eds.) Proceedings of the 2008 Conference on Current Issues in Computing and Philosophy, pp. 116–121. IOS Press, Amsterdam (2008)

Nelson, R.J.: The logic of mind. D. Reidel Pub. Co., Kluwer Boston, Dordrecht, Holland (1982)

Nelson, R.J.: Church's thesis and cognitive science. Notre Dame Journal of Formal Logic 28(4), 581–614 (1987), doi:10.1305/ndjfl/1093637649

Parikh, R.: Church's theorem and the decision problem. In: Craig, E. (ed.) Routledge Encyclopedia of Philosophy. Routledge, London (1998), http://www.rep.routledge.com/article/Y003 (retrieved)

Piccinini, G.: Computers. Pacific Philosophical Quarterly 89(1), 32–73 (2008), doi:10.1111/j.1468-0114.2008.00309.x

Post, E.L.: Finite Combinatory Processes-Formulation 1. The Journal of Symbolic Logic 1(3), 103–105 (1936), doi:10.2307/2269031

Putnam, H.: Representation and reality. The MIT Press, Cambridge (1988)

Scheutz, M.: When Physical Systems Realize Functions... Minds and Machines 9(2), 161–196 (1999), doi:10.1023/A:1008364332419

Seligman, J.: The Scope of Turing's Analysis of Effective Procedures. Minds and Machines 12(2), 203–220 (2002), doi:10.1023/A:1015638814511

Shanker, S.G.: Wittgenstein's remarks on the foundations of AI. Routledge, London (1998)

Smith, B.C.: On the origin of objects. MIT Press, Cambridge (1996)

Smith, B.C.: Is this digital after all? Managing knowledge and creativity in a digital context. Library of Congress Lecture Series (2005), http://www.loc.gov/today/cyberlc/feature_wdesc.php?rec=3381 (retrieved)

Smith, B.C.: Age of significance: Introduction (2010), http://www.ageofsignificance.org (retrieved)

Soare, R.I.: Computability and Recursion. The Bulletin of Symbolic Logic 2(3), 284–321 (1996)

Soare, R.I.: Turing oracle machines, online computing, and three displacements in computability theory. Annals of Pure and Applied Logic 160(3), 368–399 (2009), doi:10.1016/j.apal.2009.01.008

Tucker, A.B. (ed.): Computer science handbook, 2nd edn. Hall/CRC, Boca Raton, Florida (2004)

Turing, A.M.: On Computable Numbers, with an Application to the Entscheidungsproblem. Proceedings of the London Mathematical Society s2-42(1), 230–265 (1936), doi:10.1112/plms

Turing, A.M.: Systems of Logic Based on Ordinals. Proceedings of the London Mathematical Society s2-45(1), 161–228 (1939), doi:10.1112/plms/s2-45.1.161

Turing, A.M.: Computing Machinery and Intelligence. Mind 59(236), 433–460 (1950)

Von Neumann, J.: The general and logical theory of automata. In: Taub, A.H. (ed.) Collected Works, vol. V, pp. 288–328. Pergamon Press, Oxford (1961)

Wells, A.J.: Turing's Analysis of Computation and Theories of Cognitive Architecture. Cognitive Science 22(3), 269–294 (1998), doi:10.1207/s15516709cog2203_1

White, G.: Descartes Among the Robots. Minds and Machines 21(2), 179–202 (2011), doi:10.1007/s11023-011-9232-4

Chapter 4
The Triviality "Account" Examined

4.1 Introduction

In the present chapter, we examine the views of two well-known and influential philosophers concerning the ubiquitous character of concrete computation. They both share a common objective, namely, the falsification of the computationalist thesis that cognition is computational. But in order to accomplish this objective, they in effect undermine the demarcation of digital computation as a subject matter. For, by their lights, there is no fact of the matter about what digital computation is. On their view, every physical object implements any number of programs and, hence, computes every Turing-computable function. Clearly, if every physical object computes, then concrete computation ceases to be interesting in its own right.

Under the "Triviality Account" label we include the Searle-triviality Thesis and the Putnam-triviality Theorem. Searle claims that even the wall behind him can be interpreted as implementing the (now obsolete) WordStar program (1990). He offers no formal proof of this claim, but rather relies on the ill-defined concepts 'digital computation' and 'implementation'. Putnam claims that *every* physical object is a realisation of *every* abstract FSA (1988, pp. 121–125). Yet, unlike Searle, he offers a formal proof of his theorem and makes some assumptions about both the nature of the physical object in question and the type of physics invoked in the proof.

To be sure, neither the Searle-triviality thesis nor the Putnam-triviality theorem is examined here as an *account* of concrete computation *per se*. The "Triviality Account" label is used as a matter of convenience. The main objective of this chapter is to examine the repercussions of Searle and Putnam's claims, *if* they were true, for concrete digital computation as an intellectually independent subject matter. What follows from the Searle-triviality thesis and the Putnam-triviality theorem is not only *weak* pancomputationalism, that is, the thesis that *every* sufficiently complex physical system computes *some* Turing-computable function. It is rather *strong* pancomputationalism that follows, that is, the thesis that *every* sufficiently complex physical system computes *every* Turing-computable function. Nevertheless, the evidence that Searle and Putnam provide in support of their claims is shown to be weak.

N. Fresco, *Physical Computation and Cognitive Science*,
Studies in Applied Philosophy, Epistemology and Rational Ethics 12,
DOI: 10.1007/978-3-642-41375-9_4, © Springer-Verlag Berlin Heidelberg 2014

4.2 The Searle-Triviality Thesis

On Searle's view, the characterisation of a process as digitally computational is *external knower-dependent.*[1] He asserts that some alternative definitions of digital computation (for example, those suggested by Smith) may be required to emphasise some salient features of digital computation. These features include the causal relations that obtain among program states, the programmability and controllability of digital computing systems as well as their situatedness in the real world (Searle, 1990, p. 27). However, Searle claims that even given such improved and more precise definitions, the central challenge of *syntax* being essentially a knower-relative notion remains unanswered.

On his view, there is no fact of the matter about what makes a physical process, or object, computational. For digital computation is defined in terms of its syntax that is assigned externally to the computational process. The assignment of syntax is epistemically relative, rather than being *intrinsic* to the computational process. Therefore, by Searle's lights, the computational states of the physical object are not *intrinsic* to its physics. Accordingly, any *account* of concrete digital computation would be necessarily epistemically relative.

Searle's Triviality Thesis:

> For any sufficiently complex physical object O (i.e., an object with a sufficiently large number of distinguishable parts) and for any arbitrary program P, there exists an isomorphic mapping M from some subset S of the physical states of O to the formal structure of P (Searle, 1990, pp. 27–28).

How does Searle justify this thesis? For any physical object to be "sufficiently complex" it simply needs to have a sufficiently large number of distinguishable parts. A big enough wall, for instance, would do the trick (if that wall is not big enough, simply pick a bigger wall!). That wall may then be physically described at a microscopic level specifying some movement pattern of a large enough number of molecules. Further, any arbitrary program, say, WordStar, has some specification that describes its formal structure. This formal structure can in turn be implemented by any number of physical substrates as dictated by the principle of multiple realisability. "The physics [of the implementing substrate] is *irrelevant* except insofar as it admits of the assignments of 0's and 1's and of state transitions between them" (Searle, 1990, p. 26, italics added).

What remains to be seen is how M is fixed by the implementation of P. Here Searle was (mis)guided by Turing's analysis of the human computor executing an algorithm. The TM goes through the steps of P based on the rules P provides for deriving the output symbols from the input symbols. This process is causal insofar

---

[1] To be precise, Searle argues that computation is *observer*-relative. But as noted in Chapter 2, it is better to replace the 'observer' with a 'knower'. This does not weaken Searle's position regarding computation being epistemically relative to rational agents (be that humans, Martians or dolphins).

as some physical properties of O, which implements some FSA or a TM, suffice for O to follow these steps (Searle, 1990, pp. 32–33). In short, M maps the states in S onto the abstract states of P. Since, by Searle's lights, M need not be a one-to-one mapping, it suffices that only the states in S partake in the mapping from O to P. The requirement that O have a sufficiently large number of distinguishable parts supposedly guarantees that such a subset S exists. Accordingly, a wall is bound to have the right pattern of moving molecules that "mirrors" the state transitions of WordStar (or any other program specification) (Searle, 1990, pp. 27–28).

4.3 The Putnam-Triviality Theorem

In a similar vein, Putnam argues that every *ordinary open system* simultaneously realises every abstract inputless FSA (1988, p. 121). The expression 'ordinary open systems' refers to continually changing systems that are open to influences such as gravitational and electromagnetic forces. In contrast to a closed system at equilibrium that typically assumes a homogeneous state, in an open system there is a competition among various tendencies that affect the state transitions of the system. In snowflake formation, for example, there is a competition between the diffusion of water molecules towards the flake and the microscopic dynamics of crystal growth at the solid–vapour interface (Ben-Jacob & Levine, 2001). We should note that Putnam's notion of an ordinary open system accords with our present characterisation of computing systems that interact with their environment, as described in the previous chapter.

Putnam's triviality theorem was part of a carefully orchestrated attack on CTM. In particular, he argued that "a straightforward identification of mental states with [...] computationally characterized states [...] cannot be right" (Putnam, 1988, p. xii). Putnam was set to show that if the possession of some mental states were equivalent to the possession of a particular "functional organization", then it would likewise be true that the possession of some mental states is equivalent to the possession of particular behaviour dispositions (1988, p. 125). In other words, if CTM were true, it would render (the once popular) behaviourist doctrine true as well. To support his attack on CTM, Putnam identifies *computational description* with *functional organisation* and proves a theorem according to which in a sense every physical object has every functional organisation and, thus, also every *computational description* (1988, p. xv).

In order to explicate Putnam's theorem and its proof, we introduce four useful definitions followed by a formal description of the theorem.[2]

---

[2] The following definitions and Putnam's triviality theorem are adaptations of the ones proposed by Scheutz (1999, pp. 166–167).

Definition 1: a *maximal state* is "the value of all the field parameters at all the points inside the boundary of" an open system K at time t_i (Putnam, 1988, p. 122). Put otherwise, a maximal state is a complete specification of the values of *all* the relevant variables of K, inside its boundaries, at t_i.

Definition 2: a *physical state* of the system K is the set of maximal states of K for some given real-time interval (say, from 12:00 to 12:04) such that these maximal states correspond to the states of an inputless FSA δ in a single run of δ.

Definition 3: a *state type* is the union of all the legitimate[3] physical states of K corresponding to a single state of δ.[4]

Definition 4: K, as described in a physical theory P, *realises* n computational steps of δ within a given interval I of real time, if there exists a one-to-one mapping F from δ's state types onto physical state types of K and a division of I into n subintervals such that for any two states q, p of δ the following condition holds: if q→p is a transition of δ from the computational step j to $j+1$ (for $0 < j < n$) and K is in state $F(q)$ during the jth subinterval, then K transitions into state $F(p)$ in the $j+1^{th}$ subinterval.

Putnam's Triviality Theorem

There exists a theory P such that for any K, an inputless δ, $n > 0$ (a number of computational steps of δ) and for any divisible real-time interval I, K, describable in P, realises n computational steps of δ within I.

As the basis of this theorem, Putnam puts forward two principles that are crucial to his proof.

1. **The principle of continuity:** "[t]he electromagnetic and gravitational fields are continuous, except possibly at a finite or denumerably infinite set of points" (Putnam, 1988, p. 121).
2. **The principle of non-cyclical behaviour:** "[t]he system [K] is in different maximal states at different times. This principle will hold true of all [open] systems that are not shielded from electromagnetic and gravitational signals from a [natural] clock" (Putnam, 1988, p. 121).

Why are these two principles required for Putnam's proof? The principle of continuity has the modal status of a physical law of nature (Buechner, 2008, p. 147) and it assumes a theory P in *classical* physics. It states that the electrical and gravitational fields that influence K are continuous. This principle is required for

---

[3] By legitimate states Putnam includes those states that have "'something in common' [which] must [in] itself be describable at a physical, or at worst a computational, level" (1988, p. 100). Thus, "[w]e must restrict the class of allowable realizers to disjunctions of basic physical states [i.e., maximal states]" (ibid).

[4] A consequence of this definition is that there exists an isomorphic mapping that obtains between physical state types of K and δ's state types.

Putnam's proof to show that K being in a maximal state s_i at time t_1 (given the boundary conditions of K) causally determines the transition of K to a maximal state s_j at time t_2. According to the principle of non-cyclical behaviour, K is in different maximal states at different times. This principle ensures that there are *only* one-to-one mappings of the physical states of K to the computational states of δ. A natural clock continually stamps the system in such a way that K cannot have a duplicated physical state and enter the same maximal state twice.

To prove his theorem, Putnam demonstrates that *any* physical system could realise *some* arbitrary FSA (1988, pp. 122–123). To this end, he maintains that a causal relation has to be established between computational states A and B. Putnam asserts that as a result of the system being in state A and its boundary conditions the system transitions to state B. Given the state of the boundary of the system at time t_i, then it follows from the supporting lemma[5] and the principle of continuity that the inner part of the system has to change from one maximal state it was at time t_{i-1} to a distinct maximal state. "[A] mathematically omniscient being can determine from the Principle of Continuity that the system [...] must have been in [that particular maximal state]" (Putnam, 1988, p. 123), Thus, Putnam concludes, the transition from state A to state B is causal (Chrisley, 1994, p. 409).

Without loss of generality, Putnam picks an FSA that goes through the sequence of computational states $ABABABA$ in, say, a 7-minute time interval from and including 12:00 up to 12:07. Let t_1, t_2,..., t_7 be the discrete time steps corresponding to the beginning of each of the states of this FSA (i.e., $t_1 = 12:00$, $t_2 = 12:01$, ...). Let α be the physical state consisting of K's maximal states $\{s_1, s_3, s_5, s_7\}$ and β be the physical state consisting of the maximal states $\{s_2, s_4, s_6\}$. In this particular case, A will be mapped to α and B will be mapped to β in the time interval $t_1 - t_7$. This will show that K realises the given transition table during that interval. K may be either in state A or B at any given time in that interval. Since this technique of proof applies to *any* transition table of an inputless FSA, it will have proved that K can be correctly ascribed *any* transition table in the sense that there really are physical states with respect to which K realises the table ascribed (Putnam, 1988, pp. 122–123)[6].

This technique of proof, according to Putnam, can also be extended to apply to FSA *with* input and output, in other words, to finite state transducers (FST)[7]. Putnam's triviality theorem was proved for inputless FSA. If a physical object does not have input/output "organs" of the required kind, then it cannot be a model of a description that refers to a kind of FST. Further, even if the physical

---

[5] Putnam proves a supporting lemma showing that a system K' with the same spatial boundaries as K violates the principle of continuity under some appropriate conditions (1988, pp. 121–122).

[6] For a more detailed description of how Putnam's proof proceeds see, for example, Chrisley (1994).

[7] An FST is, in essence, a standard FSA that is also equipped with an output tape.

object does possess such inputs/output "organs", it may behave in a way that violates predictions, which follow from the FST description. Clearly, these inputs and outputs have some constrained realisations and one cannot arbitrarily choose some physical states to serve as their realisations. Yet, by Putnam's lights, we can find the appropriate physical states the system is in at each of the specified intervals and which stand in the appropriate causal relations to one another and to the inputs and the outputs. And hence, so he argues, the technique of proving the theorem should be similar to the one depicted above (Putnam, 1988, p. 124).

4.4   The Key Requirements for a System to Perform Digital Computation

On the triviality account, the key requirements for a physical system to perform digital computation are as follows.

- 1st Requirement: the existence of a knower.
- 2nd Requirement: the system being sufficiently complex.

The first requirement is of a *metaphysical* nature. Searle asserts that if there were no knowers, there would be no physical systems performing digital computation (1990, p. 28). For, similarly, chairs would not exist except relative to some knowers, who regarded them as such. A knower is required to assign a "computational capacity" to the physical system that exercises that capacity. The broader consequence of this view is that some physical phenomena are *knower-relative*, while others are *physically intrinsic*. Accordingly, the property of being a digital computing system, being a chair or being a glass is knower-relative. Molecular movements and gravitational forces, on the other hand, are supposedly objective, because their essential properties are *intrinsic* to their physics. Molecular movements and gravitational forces would still exist even in the absence of any knower.

By contrast, the second requirement is of a *physical* nature. Unlike Putnam, Searle does not restrict the class of permissible physical objects so as to exclude closed systems at equilibrium. He argues that for *any* computer program there is some sufficiently complex physical object and some description of this object under which it executes that program and thus performs digital computation (Searle, 1990, p. 27). But Searle also asserts that a "big enough wall" would be implementing *any* program, since it has various causal patterns of molecule movements each of which is isomorphic with the formal structure of the program implemented.

It follows then that the second requirement reduces to the existence of some arbitrary physical movement of discernible parts. These parts could be *macroscopic*, such as the holes or their absence in old-fashioned computer punched cards, or *microscopic*, such as molecules, atoms or electrons.[8] To be deemed sufficiently complex, there should be some movement of "physical parts" in the physical object that corresponds to some description of a program in terms of state transitions. According to Searle, any physical object that has a sufficiently large number of discernible parts could be said to implement any program, and, thus, to digitally compute, under the appropriate description. Because Searle does not impose any restriction on either the type or size of the physical parts in question, nor does he restrict the analysis to classical physics, it is hard to see how this view does not lead to strong pancomputationalism.

Putnam, on the other hand, only subscribes to a weak version of the first requirement and advocates a stricter version of the second requirement. Putnam's theorem suggests that any physical object (i.e., an ordinary open system) is interpretable by some knower as performing any computation. Yet, it does not follow that this interpretation is *knower-relative* in the sense advocated by Searle. For Putnam imposes sufficiently rigorous constraints (even if they are not justified) on the *conditions* of interpretation. Whether or not some specific phenomenon is knower-interpretable in some manner does not merely *depend on the knower*. For that interpretation to hold, it also *depends on the phenomenon* concerned (Chrisley, 1994, pp. 405–406). For example, many paradigmatic non-computing systems may nevertheless be interpreted as a single state FSA without input. However, arbitrarily many more non-computing systems cannot be correctly interpreted as, say, 100-state FST (which may also introduce some branching behaviour of the FST)[9].

As regards the second requirement, Putnam imposes constraints on the system concerned and the individuation of its *relevant* physical states. Unlike Searle, he does not merely settle for a large enough number of discernible parts. Putnam also assumes that the physical states of the system should be individuated by their relative *intrinsic* properties, rather than by their *sequential temporality*. This is implied by the introduction of the principle of non-cyclical behaviour that requires that the system be in different states at different times. By Putnam's lights, this principle holds true for *all systems* that are *open to external influences*. He attempts to restrict the possible choices of *relevant* physical states, which supposedly correspond to computational states, to those "'natural' states which *really do* have something in common" (Scheutz, 1999, p. 167, italics original).

Lastly, by proving the supporting lemma and imposing the two physical principles above, Putnam is set to show that the system continually transitions

---

[8] Of course, both the number and complexity (in terms of, say, thermodynamic entropy) of the parts increase when the parts are microscopic rather than macroscopic. It is, thus, more likely that for an object to realise every program simultaneously, an analysis in terms of *microscopic* parts is more appropriate.

[9] If that type of FST still seems insufficient to preclude many systems from being described as implementing them, then consider instead a CSA with 500 state-vectors and each substate of the state-vector may take 10 possible values. There is more said on CSA in the following section.

between states. Given the boundary conditions of the system at any given time interval, as well as the system being unshielded from external signals and "following" the laws of nature, the system continually evolves during the whole time interval under consideration (Putnam, 1988, p. 123). The system cannot be assigned distinct computational states A and B at t_1 and t_2 respectively in terms of their intrinsic properties, if the corresponding physical states at t_1 and t_2 cannot be distinguished from one another. Under Putnam's interpretation of ordinary open systems, if a system can be individuated into the right number of distinct states necessary to implement any FSA, then it is *sufficiently complex* (Chrisley, 1994, p. 411).[10]

4.5 Trivialisation of Computation Blocked

For Searle, the fact that TMs could be physically implemented on "just about anything" has an adverse consequence, namely, that every physical system performs digital computation (1990, pp. 25–26). Whilst Searle refers to the TM as "the standard definition of computation" (1990, p. 25), his triviality thesis seems to rely on FSA computation, for it does not mention any *memory write* operations. He takes "the class of computers [to be simply] defined syntactically in terms of the assignment of 0's and 1's". Thus, as long as the system is both sufficiently complex and properly interpreted by a knower, on Searle's view, it computes! This is probably not what Turing had in mind when he introduced the TM (the operation of the liver is also describable in terms of 0s and 1s).[11]

Searle's triviality thesis is highly questionable. Chalmers, for one, argues that the implementation relation between abstract automata and physical systems is objective and is not relative to a knower (1994, pp. 395–396, 1996, pp. 331–332). The states of Searle's wall will most likely not satisfy the *relevant* reliable state-transition rules, for this wall does not possess the required causal organisation to compute. Whilst physical systems can be easily shown to implement simple FSA with one or two internal states, the same does not hold for Chalmers' proposed CSA (1994, 1996).

---

[10] According to Ron Chrisley, "Putnam believes that *every* [...] open physical system is, in fact, arbitrarily complex" (1994, p. 411, italics added). If that is correct, then the gap between Putnam and Searle in terms of the second requirement specified above is even smaller.

[11] In fact, some authors have described the liver as a computing system (Paton et al., 1992). They have argued that the liver, in the same manner as the digital computer, is a complex system that processes data and is describable in terms of architecture, mechanism, organisation and functionality (Paton et al., 1992, p. 412). Yet, they acknowledge that in their description they have employed models and *metaphors* of parallel computing systems (Paton et al., 1992, p. 419). The question that should be asked then is whether the liver is merely metaphorically *modelled* as a computing system or whether its description *qua* a computing system does any *explanatory* work.

To reprise, a CSA is similar to an FSA except that all its states are described as state-vectors rather than as monadic states (Chalmers, 1994, pp. 393–394). Chalmers argues that simple FSA are unsatisfactory for many purposes, because of the monadic character of their states (2011, p. 328). As well, the states in most computational formalisms have a combinatorial structure. As examples, one may consider the (m-configuration, scanned symbol) combination of a TM, a cell pattern in a cellular automaton, as well as variables and registers in C or Java programs.

By Chalmers' lights, CSA are preferable to FSA for several reasons, of which only three are mentioned here. First, the conditions for implementing a CSA are much more constraining than those for implementing the corresponding FSA. The former type of automata requires the implementing system to consist in a complex causal interaction among a number of separate parts. Second, CSA reflect in a much more direct way the formal organisation of those computational formalisms listed above (i.e., TMs, cellular automata and C programs). Third, the CSA framework allows a unified account of the implementation conditions for both finite and infinite machines (Chalmers, 1996, p. 324).

Importantly, for our purposes, Chalmers argues that the conditions for implementing a given complex computation using CSA will be typically sufficiently rigorous to block arbitrary state-transition mappings à-la Searle. It would certainly be significantly more complex to find appropriate mappings for CSA than for FSA (with the same number of states overall).[12] However, the appropriate FSA can adequately mimic the behaviour of cellular automata, Pascal programs and TMs for inputs of bounded size. As mentioned in the previous chapter, a TM without memory storage, in principle, reduces to an FSA. Besides, FSA are commonly invoked in computer science when analysing programs, algorithms and digital logic circuits. Although an FSA is finite, by definition, that finiteness does not impede its adequacy as a mathematical model of computation. A very important feature of algorithms and programs is that they can generally produce infinite output by *finite* means.

Yet, while infinite CSA may be used for "translating"[13] TMs with infinite tapes, they may also be able to "compute" uncomputable functions. Indeed, if a CSA is to be suitable for transcribing a TM with an infinite tape, the state-vector of the CSA should have an infinite number of substates. Suppose that a CSA is

---

[12] Chalmers argues that the conditions for implementing a CSA are stricter than those for implementing an FSA. The state-transition rules are determined by specifying for each substate in the state-vector, a function by which its new state depends on the old overall state-vector and input vector and similarly for each element of the output vector. So, according to Chalmers, a CSA, whose state-vectors consist of 1000 substates, with each substate having 10 possible values, there will be at least 10^{1000} constraints on state transitions (2011, p. 331).

[13] "Translation" in this context is the conversion of a TM into a CSA that computes the same function that the TM does by mimicking its states.

described by an exhaustive list that includes all the possible state-transition rules of some TM by means of ordered pairs (m-configuration, scanned symbol). The result would be that the state-transition function of the CSA would need to be able to take infinitely many arguments, so that an exhaustive list would have infinitely many state-transition rules. But once the state-transition function potentially takes infinitely many arguments, it is hard to see how to prevent such a CSA from having the ability to find the solution(s) of uncomputable functions. If CSA cannot be constrained to preclude such scenarios, then the CSA framework is inadequate for modelling computation (Brown, 2003, p. 29).

It is also not so clear how TMs can be "translated" straightforwardly into CSA. The complexity of "translating" a TM into a CSA increases rapidly with the size of the TM's tape, the number of possible symbols and the number of TM's states (Brown, 2003, p. 28). Consider the simple case of the 5-state TM, which was described in the previous chapter. Since each one of the eight central squares of its tape may only contain a '0' or a '1' (the remaining two squares bearing the blank symbol are auxiliary squares for delimiting the numeral inscribed), there are $2^8 = 256$ possible tape states. The TM has five possible internal states, so we get $256 \times 5 = 1,280$ possible states. Lastly, some CSA states are also needed to indicate the position of the TM's head, which must be on only one specific square. We, thus, get a total of $1,280 \times 10 = 12,800$ distinct states the CSA can be in. This requires an exhaustive list of 12,800 transition rules, one for each state of the CSA, whereas the TM's machine table consists of only *ten* transition rules.

Be that as it may, Searle's trivialisation of computation may also be blocked without changing the "rules of the game" and replacing FSA with CSA. For one thing, Searle's triviality thesis may be refuted, since it cannot explain some computational counterfactuals, which are possible in computing systems proper. As Block points out, a computing system should allow for all the *possible* computations, which the system *could have* performed, rather than just the one it *actually* performs (2002, pp. 77–78). According to Searle, we may regard one aspect of his wall at time t_i as standing for the symbol '0' and another at time t_{i+1} as standing for '1'. So, supposedly, in the time interval between t_i and t_{i+1}, the wall computed $0 + 1 = 1$.

However, Searle's wall should also allow for all the other possible computations that could have been performed by a computing system proper. Block exemplifies this point by a 5-state FSA[14] describing the operation of an XOR gate. This example is illustrated in Figure 4.1. The numerals next to each arrow between two states represent inputs[15] and both S_4 and S_5 are accepting states with outputs. The computation of $0 + 1 = 1$ is described by the path $S_1 \rightarrow S_2 \rightarrow S_5$. The computation of $1 + 0 = 1$ is described by the path $S_1 \rightarrow S_3 \rightarrow S_5$, etc.

---

[14] Since Block makes use of the outputs of the automaton, this automaton is more accurately described as an FST.

[15] The possible inputs to this FST are (0, 0), (0, 1), (1, 0) and (1,1).

For Searle's wall to be regarded as a computing system proper, it is not enough that it has some physical states corresponding to the states '0' and '1' followed by a physical state corresponding to '1'. It should also allow for all the other *possible* computations, e.g., had the '0' input been replaced by a '1', the wall's output would have been replaced by a '0'. The burden is on Searle to show how such nontrivial isomorphism exists, particularly as he argues that a big enough wall computes *every* Turing-computable function.

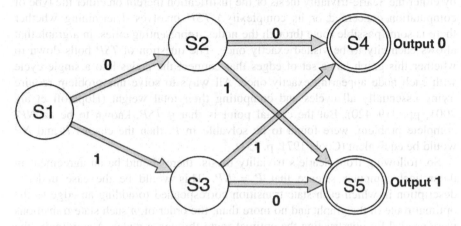

Fig. 4.1 This 5-state FST describes the operation of a XOR gate. The numerals next to each arrow represent inputs. S_4 and S_5 are accepting states with outputs.

Searle's metaphysics of computation (SM) can also give rise to some absurd consequences. For example, if we accepted SM and the theory of complexity (TC), then some unavoidable absurd consequences would follow. One surprising consequence is that there could be a knower arguing that the complexity classes *P* and *NP* are equivalent, since Searle argues that computation is knower-relative. This means that the knowledge that P=NP is produced through some computational process. Of course, it may, but *need not*, be the case. Such knowledge, for example, might be the result of an existential proof, in which case gaining this knowledge need not be computational. Nevertheless, since Searle-triviality thesis gives rise to *strong* pancomputationalism, *every* physical process, including cognitive processes, *is* deemed computational.

The question whether *P* = *NP* has been the source of disagreement in theoretical computer science for several decades and it remains unsettled. Some assert that "what we will gain from **P** = **NP** will make the whole Internet look like a footnote in history" (Fortnow, 2009, p. 80, emphasis original). Let us briefly reprise the difference between these two computational complexity classes. Familiar examples of computational problems belonging to class *P* (or *polynomial* execution time) include finding the greatest common divisor of some positive integers, the matching problem[16] and the shortest path problem[17]. Familiar examples of computational problems belonging to class *NP* (or *non-deterministic polynomial* execution time) include the Travelling Salesperson problem (*TSP*), finding a Hamiltonian cycle[18] in a graph and determining whether a mathematical statement has a "short proof" (Fortnow, 2009, p. 80).

In accordance with his triviality thesis, Searle could argue that a big enough wall could compute *TSP* in polynomial time. (For there is no restriction imposed by either the Searle-triviality thesis or the justification thereof on either the type of computation concerned or its complexity.) *TSP* involves determining whether there is some possible route through the nodes, representing cities, in a graph that allows each city to be visited exactly once. The question of *TSP* boils down to whether this graph has a set of edges that connect the nodes into a single cycle with each node appearing exactly once. All ways to solve this problem require trying essentially all cycles and computing their total weight (Hopcroft et al., 2001, pp. 419–420). But the crucial point is that *if TSP*, known to be an *NP*-complete problem, were found to be solvable in *P*, then the classes *P* and *NP* would be equivalent (Cook, 1971, p. 151).

So, following from Searle's triviality thesis, there should be an agreement in theoretical computer science that $P = NP$. This would be the case under a description in which each state transition corresponded to adding an edge to the optimal route in the graph, and no more than, the order of, *n* such state transitions were needed for constructing the optimal route through *n* nodes. Accordingly, the sequence of state transitions is upper-bounded by polynomial time. Searle's knower would simply be cognisant of what the optimal route is for a given input, for SM does not support counterfactuals about computation and there exists some arbitrary description under which the wall computes *TSP* in *P*. Indeed, it could be the case that Searle's knower would use an exponential time algorithm by trying every possible route. Nothing in the Searle-triviality thesis requires an *optimal* algorithm, but there is nothing that prevents it either. And if *TSP* turned out to be in *P*, it would follow that $P = NP$.

Searle's triviality thesis could lead to a questionable "proof" that $P = NP$. But clearly, in nontrivial counterfactual supporting computation, such a maneuver will not work. For a non-omniscient knower cannot "know" in polynomial time what the optimal route is for *any* given input. Searle's triviality thesis only requires that

---

[16] The matching problem may be exemplified as follows. Suppose that $2n$ people participate in a friendly competition in which the participants are randomly paired for a particular task. Yet, some of the participants are averse to some of the others. The problem then is either finding a set consisting of *n* pairs of participants such that each pair consists of compatible participants or deciding that such a set does not exist.

[17] The shortest path problem is the problem of finding a path between two nodes of a weighted graph (i.e., a graph with weighted edges) such that the sum of the weighted constituent edges is minimal.

[18] A Hamiltonian cycle is a cyclic path in an undirected graph that visits each node of the graph exactly once and returns to the starting node.

there exists a description under which the wall computes *TSP*. However, this consequence is absurd. If we rejected this consequence as unacceptable and accepted TC to be true, then SM would be rejected as false.

Although Putnam-triviality escapes some of the problems that Searle-triviality faces, the former still faces some serious problems concerning concrete computation. Putnam's theorem focuses more on the mathematical realm and ignores some physical constraints that unavoidably apply to computation in real-world physical systems. A central principle in thermodynamics states that a finite amount of space has finite entropy. This entropy in macroscopic open systems can be viewed as the freedom in the microdynamics for given imposed macroscopic conditions at the micro level.[19] Also, physical systems with finite entropy can only perform a finite number of operations (Penrose, 1989, pp. 391–417). It follows from these two principles that any finite physical system, such as macroscopic ordinary open systems, cannot perform infinitely many computations. So, Putnam-triviality can be resisted.

Moreover, Putnam's principles of continuity and of non-cyclical behaviour only alleviate some of the problems for trivial computations. Specifically, Putnam-triviality still faces a Kripke-Wittgenstein problem. This problem is an extension of Wittgenstein's rule-following conundrum. Saul Kripke asks what determines objectively whether one means *plus* rather than *quus* (1982, pp. 7–9). He introduces *quus* (symbolised by '$\oplus$') as the addition function (typically, known as plus) for all sums below a certain computed threshold (specifically, 57), but diverges for higher numbers. The function *quus* is defined as follows.

$$\forall x, y < 57: x \oplus y = x + y$$

$$\forall x, y \geq 57: x \oplus y = 5$$

The traditional usage of *plus* quickly becomes susceptible to arbitrarily many quus-like interpretations. So, the application of plus is no longer governed by a strictly defined rule.

Why does the Putnam-triviality face a Kripke-Wittgenstein problem in the spirit of *quus*? Physical computing systems may exhibit spatial discontinuities, but we can still determine which computation has been performed when a miscomputation occurs. Provided that the physical conditions under which the miscomputation has occurred are known, we can, generally, determine whether a physical system *S* was computing *F(x)*. We may then trace the computational path leading to the miscomputation in order to analyse it. Consequently, Kripke-Wittgenstein problems for counterfactual-supporting concrete computation may be blocked, since the miscomputation does not result in some arbitrary interpretation of *F(x)* where the computed function *changes* to *G(x)* subsequent to the physical malfunction occurring.

---

[19] This entropy is measured by the logarithm of the number of possible microscopic states for a given macro state of the system (Ben-Jacob & Levine, 2001, p. 986).

Indeed, Putnam excludes physical systems, which exhibit spatial discontinuities or chaotic behaviours, from consideration, but that exclusion does not suffice to block such Kripke-Wittgenstein problems. For there are physically possible open systems for which it is indeterminate whether they will exhibit physical discontinuities at some time t. Even if the physical conditions are known, Putnam-triviality does not allow us to predict in some possible systems whether they will exhibit such behaviours, and violate the principle of continuity. If S violated that principle, then S would not compute $F(x)$. If, on the other hand, S did not violate the principle of continuity, we would still not be able to predict whether S trivially computes $F(x)$. Thus, Putnam-triviality does not avoid Kripke-Wittgenstein problems (Buechner, 2008, p. 131). Lacking the capacity to explain the phenomenon of physical miscomputation, Putnam-triviality cannot decide whether S computes $F(x)$ rather than $G(x)$ at t. The computational-state to physical-state correspondence determines for each computational state what physical state implements it, so, on Putnam-triviality, there is no viable way of determining that a miscomputation has occurred at t in computing $F(x)$ (Buechner, 2008, p. 103).

Putnam's principle of non-cyclical behaviour is also problematic. It precludes the physical system, which realises the computation, from entering the same maximal state twice. Consider a simple TM, T, that goes into an infinite loop. T simply reads symbols on its tape and moves in the specified direction until it encounters an 'H' directing it to halt. Suppose that T's tape has three possible symbols 'L', 'R' and 'H', which stand for left, right and halt, respectively. If T encountered a sequence such as "RRLL" on its tape, it would be forced to shift between the middle "RL" positions forever. T would enter the same computational state A multiple times. This recurrence is certainly less frequent than it would seem at first blush. Any TM *might* enter the same internal state multiple times, but that state is not necessarily the *same computational state*, since a computational state can be said to also include the TM's head position and the contents of its tape. Admittedly, as stated by Putnam, "[w]ithout a computational formalism, the notion of a 'computational state' is without meaning" (1988, p. 84). So, we should consider an FSA instead.

For the system implementing the FSA not to enter the same corresponding physical state twice, it needs to have possibly an *infinite* number of physical states that implement a given computational state. An FSA, strictly, cannot write to its tape, so unlike in the case of a TM, the contents of its tape remains the same throughout the computation, thereby making the recurrence of the same computational state "easier". At least in the context of quantum physics, this is problematic. Finite space can only have a finite number of particles (whatever the minimum size of the elementary particles is). This is compatible with Max Planck's idea of indivisible quanta (1914). Putnam, of course, assumes classical physics. Besides, there is no guarantee that an open system will not enter the same state twice despite being subjected to external forces. For example, different forces can cancel each other out and the system may go back to a stable state after being perturbed (Bokulich, 2013, p. 36).

Moreover, Putnam's proof faces a problem concerning the manner in which states and state types are individuated. As already discussed in Chapter 2 in the context of the implementation criterion, it would be hard for this proof to avoid circularity in singling out the physical states that correspond to the computational ones. The physical states of any object are typically defined by the very theory that describes it. For instance, in atomic physics, the physical states of an object may be described in terms of properties of or relations among atoms (or subatomic particles). However, depending on the particular theory concerned, there may be simply *too many* physical states to choose from so as to establish an appropriate correspondence to some abstract states of, say, an FSA (Scheutz, 1999, p. 168). Putnam acknowledges that this problem indeed exists for his theorem (1988, p. 100).

Nevertheless, he offers no solace as to how to avoid the problem. Putnam discusses the problem of individuation of the *right* physical states and admits that one "must restrict the class of allowable realizers to disjunctions of basic physical states [...] which really do [...] have 'something in common'" (1988, p. 100). He points out that "this 'something in common' must itself be describable at a physical, or at worst at a computational level" (ibid). Yet, if these physical states do not really have 'something in common' at either the physical level or the computational level, then to single them out would be simply "to cheat" (ibid) for the purposes of establishing the state-to-state correspondence.

Whether the individuation criterion for physical states is defined within the physical theory or at some higher level of abstraction has repercussions for Putnam's theorem. If it can only be defined at the computational level, say, at an FSA-level description, then the whole endeavour runs the risk of circularity. For "what it is *to be* a certain computational state, *is* to be a set of physical states which are grouped together *because* they are taken to correspond to that *very* computational state" (Scheutz, 1999, p. 168, italics added).

In sum, Searle- and Putnam-like trivialisation of concrete digital computation can be blocked in various ways as we have seen above. For one thing, any computational description, which Searle or Putnam may attribute to a wall, for example, is accidental. In the former case, this description may be based on movement patterns of molecules or electrons in the wall, and in the latter on transitions between global states of the wall. However, these descriptions are both arbitrary and *ex post facto* and can hardly meet any adequate scientific standards. For Searle's triviality thesis and Putnam's triviality theorem to hold they have to be in a form of a reliable law-like generalisation. They cannot be in that form whilst describing the physical system concerned in an *arbitrary* and *ex post facto* manner.

Putnam's theorem, for one, does not support some counterfactuals that are fundamental in concrete computation. Specifically, this theorem does not support the counterfactual C expressed as "if the system were to be in state p, then it would transit into state q" (Chalmers, 1996, p. 312). What is required is not just that state p is followed by state q in some arbitrary time period as Putnam suggests. Rather, C requires that states be connected in a reliable or lawful manner and not simply by sheer coincidence. C has to obtain for *every* transition, which is specified in the FSA's transition table[20] and not only for those whose antecedent states occur in a given time period. The transition among states has to be reliable and yet open systems are susceptible to external influences where even the slightest perturbation affects the system's state. Putnam's proof is completely specific to the environmental conditions as they were during the time interval in question. What this proof is missing is showing that the relevant state combined with the laws of nature and the boundary conditions at t_i *completely determine* the next state at t_{i+1} (Chalmers, 1996, pp. 313–314).

Lastly, because Searle-triviality and Putnam-triviality are based on *ex post facto* descriptions, only the formal state transitions that are actualised in a particular run are mirrored in the structure of the physical system. It is certainly very rare that a digital computing system proper will actualise all its possible computational paths in a single run (Chalmers, 1996, p. 314). A miscomputation, for instance, will typically result in a path that does not exhaust all states that would have been actualised, had the miscomputation not occurred. If we arbitrarily singled out supposedly uninstantiated physical states to be mapped onto uninstantiated states of the corresponding FSA, why would they satisfy the C counterfactual?

In short, Searle-triviality and Putnam-triviality can at best be arbitrarily applied to computations that have *already* been *actualised*. Newton's law of universal gravitation applies to every particle in the universe at any time interval. That some arbitrary particle attracts some other particle with a force that is directly proportional to the product of their masses does *not* make this particular evidence a universal *law*. That law will remain in effect even in the absence of any knowers (including Issac Newton) to interpret the gravitation phenomenon as such. Similarly, to achieve law-like generalisation, Searle-triviality and Putnam-triviality should be modified to require that state-transitions be lawful. But that eliminates the triviality of concrete computation. If either Searle's triviality thesis or Putnam's triviality theorem had been true, then we would have been able to declare the ambitious vision *"a computer in every home"* an overwhelming success (at least) two decades ago.

References

Ben-Jacob, E., Levine, H.: The artistry of nature. Nature 409(6823), 985–986 (2001), doi:10.1038/35059178
Block, N.: Searle's arguments against cognitive science. In: Preston, J., Bishop, M. (eds.) Views into the Chinese Room: New Essays on Searle and Artificial Intelligence, pp. 70–79. Clarendon Press, Oxford (2002)

---

[20] The transition table of an FSA is similar to the machine table of a TM. It is a tabular representation of some transition function γ that takes two arguments and returns a value. The rows of the table correspond to states of the FSA and the columns correspond to its input(s). The entry for a row corresponding to state q and the column corresponding to input x is the state represented by $\gamma(q, x)$ (Hopcroft et al., 2001, p. 49).

Bokulich, P.: The Physics and Metaphysics of Computation and Cognition. In: Müller, V.C. (ed.) Philosophy and Theory of Artificial Intelligence. SAPERE, vol. 5, pp. 29–41. Springer, Heidelberg (2012), http://www.springerlink.com/index/10.1007/978-3-642-31674-6_3

Brown, C.: Implementation and indeterminacy. In: Weckert, J., Al-Saggaf, Y. (eds.) Computers and Philosophy 2003: Selected Papers from the Computers and Philosophy Conference, pp. 27–31. Australian Computer Society, Inc., Darlinghurst (2003), http://dl.acm.org/citation.cfm?id=1082145.1082150 (retrieved)

Buechner, J.: Gödel, Putnam, and functionalism: a new reading of Representation and reality. MIT Press, Cambridge (2008)

Chalmers, D.J.: On implementing a computation. Minds and Machines 4(4), 391–402 (1994), doi:10.1007/BF00974166

Chalmers, D.J.: Does a rock implement every finite-state automaton? Synthese 108(3), 309–333 (1996), doi:10.1007/BF00413692

Chalmers, D.J.: A computational foundation for the study of cognition. Journal of Cognitive Science 12(4), 323–357 (2011)

Chrisley, R.L.: Why everything doesn't realize every computation. Minds and Machines 4(4), 403–420 (1994), doi:10.1007/BF00974167

Cook, S.A.: The complexity of theorem-proving procedures. In: STOC 1971: Proceedings of the Third Annual ACM Symposium on Theory of Computing, pp. 151–158. ACM Press (1971), doi:10.1145/800157.805047

Fortnow, L.: The status of the P versus NP problem. Communications of the ACM 52(9), 78–86 (2009), doi:10.1145/1562164.1562186

Hopcroft, J.E., Motwani, R., Ullman, J.D.: Introduction to automata theory, languages, and computation. Addison-Wesley, Boston (2001)

Kripke, S.A.: Wittgenstein on rules and private language: an elementary exposition. Harvard University Press, Cambridge (1982)

Paton, R.C., Nwana, H.S., Shave, M.J.R., Bench-Capon, T.J.M.: Computing at the tissue/organ level. In: Varela, F.J., Bourgine, P. (eds.) Toward a Practice of Autonomous Systems: Proceedings of the First European Conference on Artificial Life, pp. 411–419. MIT Press, Cambridge (1992)

Penrose, R.: The emperor's new mind: concerning computers, minds, and the laws of physics. Oxford University Press, Oxford (1989)

Planck, M.: The theory of heat radiation. Maple Press, P. Blakiston's Son & Co., New York (1914)

Putnam, H.: Representation and reality. The MIT Press, Cambridge (1988)

Scheutz, M.: When Physical Systems Realize Functions... Minds and Machines 9(2), 161–196 (1999), doi:10.1023/A:1008364332419

Searle, J.R.: Is the brain a digital computer? Proceedings and Addresses of the American Philosophical Association 64, 21–37 (1990)

Chapter 5
Semantic Accounts of Computation Examined

This chapter examines some accounts according to which external semantic content is inherent to the computational process, and so any computational explanation entails content ascription to the explanandum. The goal of these accounts is clear: by ascribing semantic content to digital computing systems proper, proponents of the semantic view of computation aim to explain cognition computationally. If both digital computation and cognition *in toto* are, as they profess, semantically individuated, then they expect that cognition can be explained computationally. Three main arguments in support of this view are presented and criticised in the last section of this chapter.[1]

In the following three sections, three semantic accounts of concrete computation are expounded and criticised: the physical symbol system (PSS) account, the formal symbol manipulation (FSM) account and a reconstruction of Smith's participatory account[2]. According to the PSS account, a physical system performs digital computation when it consists of symbols and processes, which operate on these symbols that designate other entities (Newell, 1980a, p. 157). In a similar vein, according to the FSM account, a physical system performs digital computation when it processes semantically interpreted, not just interpretable, symbols (Pylyshyn, 1984, pp. 62, 72). According to the reconstruction of Smith's account of participatory computation, any physical computing system is inherently situated in its environment in a manner in which its processes extend beyond the physical boundaries of the system, which stands in semantic relations to distal states of affairs.

Given that the focus of this chapter is on the semantic view of computation, it is worth recalling our characterisation of intrinsic, extrinsic and mathematical representations in the context of concrete computation. An example of an intrinsic

---

[1] Clearly, whatever is required for a good explanation, it is more than simply that both computation and cognition be individuated semantically. Even *if* they were both semantically individuated, their individuations might nevertheless be very different.

[2] Since this is a reconstruction of Smith's account rather than one he explicitly endorses, the characterisation presented here is not strictly what makes a physical system computational, according to Smith. The reader may recall that, as discussed in Chapter 2, Smith argues that there will *never* be an adequate account of concrete computation and he certainly does not offer one.

N. Fresco, *Physical Computation and Cognitive Science*,
Studies in Applied Philosophy, Epistemology and Rational Ethics 12,
DOI: 10.1007/978-3-642-41375-9_5, © Springer-Verlag Berlin Heidelberg 2014

representation is the primitive ADD operation in conventional digital computers where the instruction "add $R1, $R2, $R3" stands for "add the values in registers *R2* and *R3* and store the result in *R1*". The semantics of such intrinsic representation is confined to the physical boundaries of the computer. The same operation also mathematically represents an addition function that takes two arguments (addends) and maps them to a single value (the sum).[3] Extrinsic representations, such as records in an EMPLOYEES_SALARIES database table, refer to symbols, data or objects that exist outside the physical boundaries of the computing system (employees and their salaries in our case). This type of representation is external-knower-dependent: a knower assigns external semantics to symbols and strings of symbols.

Whilst the three aforementioned accounts are unavoidably committed to computation being individuated by intrinsic and mathematical representations, they are also committed to computation being individuated by extrinsic representations. The burden of the present chapter is to argue that the latter commitment may be dispensed with. However, it is not only for being committed to extrinsic representations that these three accounts are ultimately found to be inadequate for explaining concrete computation. Let us turn to examine these accounts individually.

5.1 The PSS Account

5.1.1 Introduction

Allen Newell and Herbert Simon advocated physical symbol systems as being both sufficient and necessary for producing intelligent behaviour.

> **The PSS Hypothesis**: "[a] physical symbol system has the necessary and sufficient means for general intelligent action" (Newell & Simon, 1976, p. 116).

Such systems are supposedly *sufficient* for intelligence, because they can always be programmed so as to support it. Any universal system has the potential for being any other system of the same class, if so instructed. Thus, by Newell and Simon's lights, a physical symbol system, being a universal system, can become a generally intelligent system. Yet, it *need not* be one. Besides, instructability does not entail the ability to be self-instructed, so there may be no way for such a system to transform itself into one that *is* generally intelligent. Admittedly, it is the necessary condition that carries the strong implications of the PSS hypothesis. On Newell and Simon's view, any general intelligent system *contains* some physical symbol mechanism irrespective of any other structures and processes it may have. Still, it is possible to identify in the system concerned what serves as symbols and

---

[3] Note that the range of a function is finite in a physical computing system, whereas mathematical functions can have infinite ranges.

symbolic expressions as well as the processes that operate on them (Newell, 1980a, pp. 170–171).

The ensuing discussion shall be mostly restricted to PSS as an account of concrete digital computation *proper*. According to this account, digital computing systems are physical symbol systems containing sets of interpretable and combinable entities (i.e., symbols and symbolic expressions) and a set of processes that operate on these entities by generating, copying, modifying, combining and destroying them according to instructions. These symbolic entities are physical patterns (i.e., tokens) that can occur as components of symbol structures. A physical symbol system is situated in a world of objects that is wider than just these symbolic expressions (Newell & Simon, 1976, p. 116). Whilst the PSS account resembles the FSM account, as is shown below, there are also some differences that cannot be easily dismissed.

Moreover, Newell and Simon maintained that a physical symbol system is an instance of a UTM (1976, p. 117). On their view, UTMs "always contain within them a particular notion of symbol and symbolic behavior" (Newell, 1980a, p. 155). The problem-solving capacity of a physical symbol system is exercised by producing and progressively modifying symbol structures until a solution structure is produced, typically, by means of a heuristic search (Newell & Simon, 1976, p. 120). There are two basic aspects to each search, namely, its object (i.e., what is being searched for) and its scope (i.e., the set of objects through which the search is conducted). In computing systems, each of these aspects has to be made explicit in terms of specific structures and processes, since the system cannot search for an object that it cannot recognise (Haugeland, 1985, p. 177).

Whilst UTMs and some other computing systems *may* solve some problems by using a heuristic search, it is not clear that all conventional digital computing systems do so. Physical symbol systems as instantiations of UTMs perform heuristic search as a means of overcoming their limited processing resources. They can execute only a finite number of processes in a finite number of steps and over a finite interval of time (Newell & Simon, 1976, p. 120). Heuristic search processes involve the generation, composition and transformation of symbolic structures until the specified conditions for a solution are met (Clark, 2001, p. 33).

But not every computational problem has to be solved using a heuristic search. Newell and Simon described problem-solving search as generating branching trees of partial solution possibilities (1976, p. 122). These branching trees may grow to thousands or even millions of branches before yielding a solution to the computational problem. If from each symbolic expression the search produces B new tree branches are produced, then the tree will grow as B^D (where D stands for the depth of the tree). Yet, many algorithms are not based on such a search procedure, but rather on a finite sequence of instructions for solving a particular problem. Consider, for example, a UTM that simulates the special-purpose TM introduced in Chapter 3 for computing the successor function. Such a UTM can search for occurrences of symbols and replace some with others without producing branching search trees.

Physical symbol systems are describable at least at two levels: the symbol level and the physical level. Newell asserted that symbol structures and operators[4] on these structures at the symbol or program level are realisable at the physical level (1980a, p. 156). The paradigmatic architecture that supports the realisation of symbol systems consists of a memory, a set of operators, a control, as well as input and output devices (Newell, 1980a, p. 142). The input and output devices are receptors and effectors, respectively (Newell & Simon, 1972, p. 20).[5] According to Newell, there is also another level, the knowledge level, which is distinct from the other two and located "above" the symbol level (1980b). This level is characterised by knowledge as a potential for generating action relative to the goals of the system. "Knowledge serves as the specification of what a symbol structure [at the symbol level] should be able to do" (Newell, 1980b, p. 7).

We conclude this section by briefly elaborating what the basic constituents of a physical symbol system are. The long-term memory contains symbol structures (i.e., symbolic expressions), which "vary in number and content over time" (Newell, 1980a, p. 142). The control component (or *processor* in (Newell & Simon, 1972, p. 20)) governs the behaviour of the system. The control component consists of a fixed set of elementary rules, a short-term working memory, which holds the input and output symbolic expressions, and an interpreter for determining the sequence of operations to be executed as a function of the input symbolic expression. The system's inputs are "the objects in certain locations" and its outputs are objects that are either modified or newly generated (Newell, 1980a, p. 142). Further, the operators are mechanisms that take some symbolic expressions as inputs and produce some symbolic expressions as outputs (plus possibly some other effects) as a result. The internal state of the system concerned is the combination of the system's memory state and the system's control state.

5.1.2 The Key Requirements Implied by the PSS Account

The PSS account identifies seven key requirements for a physical system to perform digital computation.

- 1st Requirement: containing a set of symbols and a set of processes that operate on the symbols and produce through time an evolving collection of symbolic expressions.
- 2nd Requirement: having the capacity to distinguish between some symbolic expressions as data and others as programs.

---

[4] Operators, such as '+', '-', or 'copy', are symbols or symbolic expressions that have an external semantics built into them (Newell, 1980a, p. 159).

[5] Strictly, in Newell and Simon (1972) the system in question is an *information processing system* (IPS). However, Newell later acknowledged that IPS were the systems at the basis of the PSS hypothesis (1980a, p. 141). So, an IPS and a physical symbol system are deemed equivalent.

- 3rd Requirement: having a stable memory for storing generated symbolic expressions until they are explicitly modified or deleted.
- 4th Requirement: having the capacity to handle an unbounded number of symbolic expressions and to realise the maximal class of input/output functions using these expressions.
- 5th Requirement: having the capacity to either affect a designated object or behave in ways that are dependent on that object.
- 6st Requirement: having the capacity to interpret a symbolic expression, if it designates some process, and execute that designated process.
- 7th Requirement: supporting all the symbolic expressions that designate the processes that the system can execute.

The first requirement is the system containing a set of symbols and a set of processes that operate on the symbols, and thereby temporally producing an evolving collection of symbolic expressions. At any given time, the system contains a collection of symbolic expressions and processes operating on them to produce other expressions. The possible operations are the generation, modification, reproduction and destruction of symbolic expressions (Newell & Simon, 1976, p. 116). These operations are performed by the control component that transforms *internal* symbolic expressions in accordance with finite basic rules. In other words, the system executes computer programs that operate on data structures (Bickhard & Terveen, 1995, p. 92).

The second requirement is the system having the capacity to distinguish between some symbolic expressions as data and others as programs (Newell, 1980a, p. 166). This is indeed a property of all UTMs that need to recognise some symbolic expressions as data (the *input* to the simulated special-purpose TM) when generating or modifying them at time T_i and then interpret the same or other expressions as programs (e.g., the instructions table of the simulated special-purpose TM) at another time T_j. This binary distinction is purely contextual, since the symbolic expression itself is the same whether it is data or an instruction to be executed. Newell specified several ways for implementing this capacity, such as having a *program* memory as distinct from the *data* memory with a transfer operation to activate the program or marking the program so that the control component does not interpret it and then removing the mark to activate the program (1980a, p. 166). The concept of *universality* is clearly fundamental to all physical symbol systems.

Nevertheless, Newell and Simon acknowledged several limitations on the universality of physical symbol systems. First, "[a]ny machine is a prisoner of its input and output domains" (Newell, 1980a, p. 148). Computing systems are situated in the real world, but their contact with it is limited to the transducers they are equipped with. This limitation is alleviated from a theoretical perspective by only considering abstract inputs and outputs. The ability to generate any computable function between two given domains enables, in principle, the generation of functions between the corresponding transducers acting as input and output channels. Thus, the universality of any physical symbol system is always relative to its input and output channels. Second, real world computing systems are subject to physical limits, such as the speed and reliability of components, available memory space and the energy available. So, the universality of the computing system is relative to those physical implementation limits.

Third, and perhaps most important, there are too many ways to instruct a machine how to behave. There is an arbitrary large number of *different* algorithms for solving the *same* computational problem. Also, there appears to be no way that a concrete universal system can literally behave according to *any* input-output function, if the time over which the behaviour is to occur is indefinitely extended (Newell, 1980a, p. 149). This was the import of the discovery of uncomputable functions. The notion of universality is also relative to a given class of computing systems. For any class of computing systems that is defined by some way of describing their operational structure, a system of that class is *universal*, if it can behave as any system of *that class* (Newell, 1980a, pp. 149–150).

The third requirement is the system having a stable memory for storing generated symbolic expressions until they are explicitly modified or deleted (Newell & Simon, 1976, p. 116). This requirement stems from the coupling of read/write operations in computing systems. Each of these two operation types requires its counterpart to be productive in affecting the system's behaviour. A read-type operation only retrieves expressions that were written to memory and persisted. Conversely, writing to memory expressions that are never subsequently read is a redundant operation (Newell, 1980a, p. 163). When the system works to solve a specific problem, it accesses its long-term memory until all the relevant instructions have been executed. This mode of operation yields a transfer of all the symbolic expressions required for the solution of the problem into the short-term working memory of the system's control component (Clark, 2001, p. 31).

The fourth requirement is the system having the capacity to handle an unbounded number of symbolic expressions and to realise the maximal class of input/output functions using these expressions. This requirement is weaker than requiring unbounded memory (the reader may recall the *infinite* tape of an idealised TM). The structural requirements for universality are not dependent on unbounded memory. Rather, they are dependent on the system's capacity for handling an unbounded number of (token) symbolic expressions (Newell & Simon, 1976, p. 116).[6] Central to the universality of physical symbol systems is their flexibility of behaviour. However, it is not sufficient that the system can produce any output behaviour. Rather, this behaviour has to be responsive to the input. Thus, a universal system is one that can produce some arbitrary input-output function relative to other systems of the same class (Newell, 1980a, p. 147). That is, the system should be able to produce any dependence of output on input.

---

[6] The reader should recall from the discussion in Chapter 3 that if a computing system used *infinitely many* symbol *types*, then some of those symbols might be too similar to the others to be reliably distinguished in a finite time.

The fifth requirement is the system having the capacity to either affect a designated object or behave in ways that are dependent on that object. "A physical symbol system [...] exists in a world of objects wider than just these symbolic expressions themselves" (Newell & Simon, 1976, p. 116). So that the manipulated symbols be *meaningful* and not just "empty syntactic shells", the physical symbol system has to be located in and interact with an environment rich in "real-world items and events" (Clark, 2001, p. 28). Relative to events and objects in its real-world environment, a symbolic expression in the system has the power to *designate* an object. That is, the symbolic expression can, but need not, be an *extrinsic* representation, in our terminology.

An entity, that is, a symbol, x designates (i.e., is about or stands for) an entity, such as an object or a symbol, Y relative to a process P, if when x is P's input, P's behaviour depends on Y. Designation is grounded in the physical behaviour of P and its action can also be "at a distance" (Newell, 1980a, p. 156). That is, P behaves as though some input, represented by x, whose source is some distal object or event, has an affect on P. This "action at a distance" is accomplished by a mechanism of access realised in physical computing systems. In a paradigmatic physical symbol system, this access mechanism is implemented as an *assign* operator, which establishes a basic relationship between a symbol and the entity to which it is assigned (Newell, 1980a, p. 144). Any of the system's operators or its control component that has access to an occurrence of this symbol has access to the assigned entity. "Access to an operator implies access to its inputs, outputs, and evocation mechanism" (ibid). If a symbolic expression can be generated at time T_i that is dependent on an entity in some way, processes can exist in the system that at time T_{i+1}, take that expression as input and behave in a way dependent on that entity (Newell, 1980a, pp. 156–157).

The sixth requirement is the system having the capacity to interpret a symbolic expression, if it designates some process, and execute that designated process. Interpretation is defined as the act of accepting a symbolic expression that designates a process as input and then executing that process (Newell, 1980a, pp. 158–159). This is similar to the process of indirectly executing computer programs by an interpreter program. The interpreter reads a symbolic expression as input and if it is recognised as a program (or a set of instructions), rather than a data structure (recall the second requirement), then it is executed. This capacity is necessary to allow the flexibility of universal computing systems to generate expressions for their own behaviour and produce that very behaviour. The total processes in the computing system can be decomposed to the basic structure of (control + (operators + data)) that is typical of the imperative programming paradigm. The control component continually brings together operators and data to yield the desired behaviour.

The seventh requirement is the system supporting all the symbolic expressions that designate the processes it can execute (Newell & Simon, 1976, p. 116). This requirement is necessary to support the full plasticity of behaviour of universal computing systems. Symbols that designate the operators of the system are essential and no number of symbols for expressions or roles can replace them.

"These are the symbols that have an external semantics wired into them" (Newell, 1980a, p. 159) (however that is accomplished). The number of such symbols varies between different instances of symbol systems, but it cannot be reduced to zero.[7]

5.1.3 The PSS Account Evaluated

Spoiler alert: the PSS account is inadequate for explaining concrete digital computation proper. Let us observe why. For one thing, it is unacceptably narrow, as it excludes too many paradigmatic digital computing systems. It, thus, fails to meet the dichotomy criterion. The PSS account certainly excludes systems such as lookup tables and digital thermostats, for they clearly fail to meet *all* the requirements above. Whilst they are not paradigmatic digital computing systems, discrete connectionist networks, for example, are ruled out by the sixth requirement. For they inherently lack a single control unit that continually brings together operators and data to yield some desired behaviour. Further, not only does the PSS account reduce digital computation at large to classic symbolic computation, but it also restricts the class of symbolic computation. For the PSS account classify all physical symbol systems as *universal* machines (or instances of UTMs). If we accept that a physical system is computational only provided that it is universal, then even physical instances of special-purpose TMs are not deemed computational.

The PSS account rules out other exemplars of digital computing systems as non-computational. The building blocks of conventional digital computers, such as shift registers or multiplexers, are clearly not universal and cannot simulate systems of the same class, thereby violating the second requirement above. We cannot build a universal shift register that simulates any other shift register. One might argue that these examples are not digital computing systems proper and so the problem lies elsewhere. However, even physical instances of FSA are ruled out by this account as non-computational. FSA cannot write to memory storage, thereby violating the third requirement above. Besides, we cannot in principle build a universal FSA that can simulate any other FSA. If (physical instances of) FSA are *not* digital computing systems proper, many books in theoretical computer science will need revision very soon.

Moreover, the PSS account excludes other early-day digital computers. One such example is Konrad Zuse's Z1 mechanical computer in which instructions were read from a punched 35mm film stock. Another example is the Atanasoff–Berry computer, which was designed only to solve systems of linear equations. Similarly, the PSS account excludes the two earliest large-scale electronic digital computers: the 1943 British Colossus and the 1945 American ENIAC. Both these

---

[7] It is not clear why it is necessary that *every* such expression exist. There could be a mismatch between the set of all functions and the set of all expressions representing them. For instance, some functions could be the serial invocation of several expressions (themselves representing other functions).

computers did not store programs in memory and required manual modification of some of the machine's wiring, rerouting cables and setting switches to perform new tasks (Copeland, 2004, p. 3). Therefore, they violate the fourth requirement above and are, thus, ruled out as non-computational. The resulting class of computing systems denoted by the PSS account is intolerably narrow and it is unclear whether the requirement(s) for universality may be dropped without drastically altering the PSS account.[8]

The PSS account also fails to meet the taxonomy criterion for similar reasons. To start with, it does not distinguish the computational capacities of most computing components and systems. For the PSS account excludes them from consideration altogether. Basic logic gates and more complex circuits (e.g., half- and full-adders, flip-flops and shift registers) are not classified as computational. Similarly, physical instantiations of special-purpose TMs and FSA are excluded from the class of digital computing systems. If these systems (and others) are not computational to begin with, they cannot be taxonomised based on their computational power. To avoid any doubt, neither Newell nor Simon denied that logic gates and logic circuits are the building blocks of conventional digital computers, but these are certainly not physical symbol systems that meet the aforementioned requirements.

The PSS account seems to fail the implementation criterion as well. One contributing factor to this failure is the *designation* requirement. Newell's description of the physical symbol system resembles that of conventional digital computers (Newell, 1980a, p. 173). At the bottom level, physical devices are described in electronic terms. Above this level he lists the circuit level, which is describable in terms of electrical currents and voltages. The next level above is the logic level, in which various logical functions occur when bits pass through functional units of the computing system. The next level is the program level, which contains symbols (or variables), data structures, addresses and programs. Newell described the top level as "the level of gross anatomy", where data flow along channels, called links and switches, and are then stored or processed by processors (or control components) and transducers (ibid).

Yet, as admitted by Newell, "the instantiation for physical symbol systems is still highly special, and much is missing [in their description]: *designation of external entities, wider ranges of interpretive activity*, and so on" (1980a, p. 168, italics added). The fifth requirement above indicates that processes in computing systems are *causally* affected by the semantics of symbols. The behaviour of some process P (with x as its input) depends on a potentially distal entity, which is designated by x.

---

[8] Strictly, the PSS account even rules out standard physical instantiations of UTMs, since UTMs can read/write/erase only *single* symbols located at a *specific* memory location at any given time. Physical symbol systems, on the other hand, can read, write and erase *structured symbolic expressions* in their entirety. Nevertheless, physical symbol systems are not more computationally powerful than UTMs.

Internal designation is clearly unproblematic in computing systems and, in our terminology, can be described as intrinsic representation. In relation to the seventh requirement, Newell wrote that "[t]he term *interpretation* is taken [...] in the narrow sense it has acquired in computer science" (1980a, p. 158). But he also added that "[some] symbols [...] have an *external semantics* wired into them" (Newell, 1980a, p. 159, italics added). The specification of external designation is vague at best, as it leaves the ways in which a computational process depends on some external entity underdetermined.

Newell and Simon *might* have taken it for granted that symbols *symbolise* by definition and so they have not explicated how external semantics gets "wired into them". If, indeed, external semantics is required for computation, then this gap is simply too big to remain unexplicated. Internal access to symbols and expressions in conventional digital computers is a primitive in the computer's architecture (e.g., for memory retrieval). But there is no similar primitive for accessing the external environment (Bickhard & Terveen, 1995, pp. 93–94). Of course, it does not follow that there is no room for *derivative* semantics, only that it is not built into the architecture of the computer. The PSS account is committed to manipulable symbols not just as *internal states* of the system that support a systematic interpretation, but they also partake in processes of copying, combining, shuffling and destroying them (Clark, 2001, p. 29). Nevertheless, how designation of external entities is physically implemented remains a mystery.

Moreover, the PSS account does not explicate how symbol transformations are mapped at lower levels of implementation. At the appropriate level of abstraction, a program executed on a conventional digital computer may be described as the manipulation of interpretable symbols. The internal working of the digital computer is abstracted away by means of a program operating on data structures (Bickhard & Terveen, 1995, p. 92). The PSS account focuses on the processing of internal "records", rather than on the influence of the relevant physical states on this processing. For the notion of 'processing' does not depend in any essential way on *action* or *interaction*. In general, processing is merely described as manipulations of internal "records". The PSS account defines 'process' in a manner that *presupposes* issues of representation. Processes operate on *symbols*, but how representation emerges out of some process remains underdetermined. The functional influence of the implementing system's internal states on processing and the limits of what internal states can do, or be, are not reflected by the PSS account (Bickhard & Terveen, 1995, p. 93).

The PSS account does not explain miscomputation at any great level of detail, but it does not ignore this important aspect of concrete computation either. For instance, Newell acknowledged that "the phrase *under normal conditions* is essential to the [...] characterization [of physical symbol systems]" (1980a, p. 175, italics original). He further elaborated that all kinds of error that occur at lower levels of the computing system typically propagate through to higher levels (e.g., the symbol level) and yield a behaviour that is revealing of the errors in the underlying structures. By simply forcing a system against the physical limits of its architecture one may reveal the details of the implementing technology.

Elsewhere, Newell and Simon asserted that some errors may occur as the result of interference among symbols in the memory or in some other time-dependent process (1972, pp. 795–796). Nevertheless, the details of how computation is distinguished from miscomputation in terms of processed physical symbols remain unexplicated.

In sum, as an account of concrete digital computation proper, the PSS account is inadequate. This account was motivated by the view that a digital computer and a human cognitive agent share much in common, and that commonality makes them both intelligent systems. In that regard, the PSS hypothesis was at the centre stage of artificial intelligence for several decades. Also, as an account of the internal workings of a *programmable stored-program universal* digital computer (barring any extrinsic representation), the PSS account is both comprehensive and insightful. But it is inadequate as an account of concrete computation *at large*. It, arguably, meets the conceptual criterion and the program execution criterion. As regards the former, this account can pay off any conceptual debts such as 'data', 'algorithm', 'program' 'compilation', 'interpretation' and 'architecture'. However, as regards the latter, in *equating* digital computation with physical symbol systems that execute programs, the PSS account classifies too many paradigmatic computational systems as non-computational.

5.2 The FSM Account

5.2.1 Introduction

According to the FSM account, digital computing systems are formal symbol manipulators. Formal symbol manipulation has been taken to be virtually synonymous with digital computation. That is, the assertion that digital computation is a formal manipulation of symbols is considered a tautology, rather than a claim, whose truth-value has to be established empirically (Smith, forthcoming). On the FSM account, digital computing systems manipulate symbol tokens, which themselves are representations of the subject matter the computation is about, in accordance with some purely formal principles (Scheutz, 2002, p. 13). Although these manipulated symbols, supposedly, have both semantic and syntactic properties, only the latter are causally efficacious.

The chief proponents of the FSM account are Fodor (1975), Pylyshyn (1984, 1989) and Haugeland (1985). Fodor asserts that "[digital] computations just are processes in which representations have their causal consequences in virtue of their form" (1980, p. 68). Haugeland's well-known formalist's motto stated that "if you take care of the syntax, the semantics will take care of itself" (1985, p. 106). A digital computing system, as an interpreted automatic formal system, "takes care of the syntax" (Haugeland, 1985, pp. 258–259, fn. 6). According to Pylyshyn, "[the] 'meaningless' manipulation of symbols [...] is just what [digital] computers are good at" (1989, p. 7) and this "is a good reason why computers can be described as processing *knowledge*" (1989, p. 6, italics added). In short, classical digital computers are viewed as representation-manipulating systems, which receive representations as input, manipulate them according to formal rules based solely on their shapes, or syntax, and produce representations as output.

However, the seemingly innocent description of digital computing systems as formal symbol manipulators is problematic and requires some unpacking before any substantial analysis can start. First, on the FSM account, digital computation is an intentional phenomenon, as the following statement implies. "A computational process is one whose behavior is viewed as depending on the *representational* or *semantic* content of its states" (Pylyshyn, 1984, p. 74, italics added). Yet, a formality condition is imposed on the operation of digital computing systems as observed by Fodor. "[A computing] device satisfies the formality condition; viz., it has access only to formal (non-semantic) properties of the representations that it manipulates" (Fodor, 1980, p. 63). In imposing the formality condition, the computing system, supposedly, avoids any difficult questions about semantics (for example, what establishes the conditions for semantic success?) (Smith, forthcoming).

Whilst Turing's account is at the very least ambivalent about taking computation to be an intentional phenomenon, the FSM account is committed to this view. Whatever this intentionality is, whether original or derivative, internal or external, the FSM account (as well as the PSS account, for that matter) would not have otherwise made 'symbol' a fundamental ingredient of its main thesis about computation. Still, interestingly, it places an equal burden on formality.

Second, it is often assumed that symbols, which are formally manipulated, need to be explicitly represented, but this immediately renders the FSM account too narrow (Smith, forthcoming). Mathematical and logical proofs are paradigmatic cases of a formal manipulation of explicitly represented syntactic formulas. Yet, insisting on the explicit encoding of symbols and symbolic expressions, rules out other possible digital computing systems, such as discrete connectionist networks, as non-computational. For connectionist networks do not inherently invoke explicit representations in the course of their operation. If connectionist networks do not invoke explicit representations but are computational nonetheless, then, on the FSM account, they are ruled out unjustifiably. This suggests that both *explicitness* and *implicitness* are the wrong ingredients to be included in a foundational account of computation.

Third, the FSM account may be interpreted in three different ways that need not extend to the same class of digital computing systems (Smith, forthcoming). On one interpretation, the FSM account makes a *positive* claim regarding computing systems, namely, that their operations proceed solely in virtue of the syntactic properties of manipulable symbols. The syntactic (i.e., shape-based or something similar) properties of those symbols *are* the *effective* ones. For example, if a digital computing system produces '1100' as the result of '1000 + 0100', then, according to the "positive interpretation", that transition will have occurred formally solely in virtue of the syntactic properties of '1000' and '0100'. On a second interpretation, the FSM account makes a *negative* claim regarding computing systems, namely, that their operations proceed independently of any semantics of the manipulable symbols (i.e., irrespective of any semantical predicates such as *truth*, *meaning* or *reference*). Yet, on a third interpretation, which may be dubbed the "non-semantical interpretation", the syntactic properties and the semantic properties of the symbols *do not overlap.*

The problem then is whether the FSM account commits to one or more of these three possible interpretations and what that commitment entails in terms of the resulting class of digital computing systems denoted by this account. Whilst these three possible interpretations *may* result in the very same class of digital computing systems, they *need not* do so necessarily. The simple course of action here would have been to restrict our analysis to the interpretation committed to by the proponents of the FSM account, if, say, Fodor, had not committed to all three simultaneously (deliberately or not). In support of the "positive interpretation", Fodor asserts that "formal operations apply in terms of the, as it were, 'shapes' of their objects in their domains" (1980, p. 64). In the same paragraph, in support of the "negative interpretation", he asserts that "formal operations are the ones that are specified without reference to such semantic properties of representations as, for example, truth, reference, and meaning" (ibid). And lastly, in support of the non-semantical interpretation, he adds that "[w]hat makes syntactic operations a species of *formal* operations is that *being syntactic is* a way of *not being semantic*" (Fodor, 1980, p. 64, italics added).

As regards the second problem mentioned above, that is, the FSM account insisting on explicit encoding of manipulable symbols, Fodor does not link *explicitness* with *formality*. He claims that in equating formal operations with those that apply effectively, one identifies formality with explicitness (Fodor, 1980, p. 72, fn. 4). By his lights, there is no good reason for insisting that 'formal' means both 'syntactic' and 'explicit'. By accepting this claim, only three possible options remain for the FSM account in that regard. First, explicitness is not part of the FSM account as a constraining property of digital computing systems. Second, explicitness is inherently linked with symbols, i.e., any manipulable symbol has to be explicitly encoded when operated upon by the computing system. Third, explicitness is inherently linked with the manipulation of symbols. The last option seems patently wrong. The manipulation of "stuff" can proceed, perhaps less efficiently, even if it is implicitly encoded. If that is right, there only remains the second option as shown below.

To allow the analysis to proceed, let us restrict the FSM account here to the version supported by Fodor, Pylyshyn and Haugeland. Having said that, a caveat is in order. There are subtle differences among their views that, strictly, lead to different versions of the FSM account. But despite those subtleties, the moral at the end remains the same regarding the adequacy of FSM as an account of concrete digital computation. A digital computing system, according to this account, is a physical automatic formal system with two main properties. The first property is that some of its states, or parts, are identified as tokens (i.e., symbols) of the system. The second property is that under normal circumstances, the system automatically manipulates these symbols according to the rules of that system (Haugeland, 1985, p. 76).

Still, not all the functions of the computing system need to be encoded explicitly. What needs to be *explicit* in those systems is not the program, but the manipulable *symbols*. The FSM account is, therefore, committed to the second option above. The rules of manipulation can be *either* implicit or explicit.[9] The UTM is rule-explicit about the special-purpose TM it simulates, for the instructions of the latter are explicitly represented (or encoded) on the tape of the former. However, the UTM's own table of instructions is *hard-wired* and those rules (or instructions) may, thus, be deemed implicit (Fodor & Pylyshyn, 1988, pp. 60–61).

5.2.2   The Key Requirements Implied by the FSM Account

The FSM account, according to Fodor, Pylyshyn and Haugeland, identifies five key requirements for a physical system to perform digital computation.[10]

- 1st Requirement: operating using internally represented rule-governed transformations of interpretable symbolic expressions.
- 2nd Requirement: the abstract states of the system have to correspond to equivalence classes of physical states such that their members are indistinguishable from the point of view of their function.
- 3rd Requirement: supporting an arbitrarily large number of symbolic expressions.
- 4th Requirement: having the capacity to capitalise on the compositional nature of symbolic expressions as determined by the constituent expressions and the rules used to combine them.
- 5th Requirement: having an accessible memory.

The first requirement is the system operating using internally represented rule-governed transformations of interpretable symbolic expressions. According to Haugeland, a digital computing system is "[a]n automatic formal system [....] which automatically manipulates the tokens of some formal system according to the rules of that system" (1981, p. 10). By following the formal rules of transformation operating on symbolic expressions, the semantically interpreted expressions need to continue to "'make reasonable sense,' in the contexts in which they're made" (Haugeland, 1985, p. 106).

---

[9] In fairness to Haugeland, he claimed that automatic formal systems "proceed according to explicit rules" (1981, p. 17). However, he qualified that claim by adding that the rules should be "explicit, at least, in the program of the virtual machine" (ibid). This qualification suggests that, by his lights, rules are explicit in program-controlled digital computing systems.

[10] These five requirements seem to be advocated by Fodor, Pylyshyn and Haugeland. Yet, as discussed below, there is at least one more requirement that is only alluded to by Pylyshyn.

Furthermore, in being a representational, or an intentional, system – the regularities in the operation of the computing system are fully explicable by referring only to the content of its representations.[11] The transformation rules need to have the property that in applying to a specific symbolic code or state they "appear to depend on what it is a code *for* [...], they must 'respect the semantic interpretation'" (Pylyshyn, 1984, p. 66, italics original). These regularities are *rule-governed*, rather than *law-like*. So, at a meta-theoretic level, any explanation of a computational process needs to make reference to what is represented by the semantically interpreted computational states and rules, rather than just to causal state transitions (Pylyshyn, 1984, p. 57).

Moreover, on the FSM account, it is a key property of computing systems that semantic interpretations of their states remain *consistent* despite the state transitions of the system. Since computational processes follow a particular set of semantically interpretable rules, semantic interpretations of computational states cannot be given capriciously (Pylyshyn, 1984, p. 58), though they need not be unique. This is analogous to rules in formal logic, such as Existential Generalisation and Universal Instantiation, that apply to formulas in virtue of their *syntactic* form, yet, their salient property is *semantical* in that they are truth preserving (Fodor & Pylyshyn, 1988, p. 29).

The second requirement is the abstract states of the system corresponding to equivalence classes of physical states, such that their members are indistinguishable from the point of view of their function (Pylyshyn, 1984, p. 56). There exists a basic mapping from atomic symbols to relatively elementary physical states, and a mapping specification of the structure of complex symbolic expressions onto the structure of relatively complex physical states. There also exists a mapping from the physical states of the computing system to its abstract states (in Fodor's words "formulae in a computing language") that preserves the semantic relations among the abstract states (Fodor, 1975, p. 73). "Token machine states are [...] interpretable as tokens of the formulae" (Fodor, 1975, p. 67). The physical constitution of the system will result in "run[ning] through that sequence of [physical] states" *only if* the corresponding sequence of abstract states "constitutes a proof of [the last abstract state]" (ibid).[12] "[I]n its normal operation, [a digital computing system] automatically manipulates [... its states] in accord with the rules of that [system ...] like a magical pencil that writes out formally correct logical derivations" (Haugeland, 1997, p. 11).

The third requirement is the system supporting an arbitrarily large number of symbolic expressions (Pylyshyn, 1984, p. 62). Digital computing systems can manipulate arbitrarily many "tokens" in any specifiable manner whatever, as long as those tokens are arranged to be symbols (Haugeland, 1985, p. 4). The architecture of conventional digital computing systems "requires that there be

---

[11] This is where it is crucial to distinguish intrinsic from extrinsic representations.

[12] In his description of the physical and computational, or abstract, state-transitions, Fodor likens the computational state-transitions to a logical proof of some formulae in the appropriate computing language.

distinct symbolic expressions for each state of affairs it can represent" (Fodor & Pylyshyn, 1988, p. 57). This raises the question of how a computing system may support an arbitrary large number of expressions with only a finite number of transformation rules. For a fixed number of expressions some sort of a lookup table can be implemented. However, this is not possible for an arbitrarily large number of symbolic expressions (Pylyshyn, 1984, pp. 61–62). Instead, this apparent "gap" is bridged by the fourth requirement.

The fourth requirement is the system having the capacity to capitalise on the compositional nature of expressions as determined by the constituent expressions and the rules used to combine them. By supporting simple rules that operate on basic individual symbols the system can generate an arbitrarily large number of symbolic expressions. Complex expressions are realised and transformed by means of instantiating constituent expressions of representations (ibid).

The semantics of a composite symbolic expression is determined by the semantics of its constituents in a consistent way and in digital computing systems there is a structural (e.g. part/whole) relation that holds between them (Fodor & Pylyshyn, 1988, p. 16). For instance, the semantics of 'the daisies in the vase on the table by the door' is determined by the semantics of 'the daisies in the vase on the table', which is determined by 'the daisies in the vase' and so forth. Most of the symbolic expressions in computing systems viewed as interpreted automatic formal systems are complexes, whose semantics is determined by their systematic composition (Haugeland, 1985, p. 96). "[I]t is characteristic of Classical [digital computing] systems [...] to exploit arrays of symbols some of which are atomic [...] but indefinitely many of which have other symbols as syntactic and semantic parts" (Fodor & Pylyshyn, 1988, p. 16).[13]

Implicitly, the fifth requirement is the system having an accessible memory. As in the case of a TM, a computing system needs to have a memory that allows for the reading *and writing* of symbolic expressions (Fodor, 1980, p. 65). This memory may consist of a running tape, a set of registers or any other storage medium (Pylyshyn, 1989, p. 56). The memory's capacity, its organisation and the means of accessing it are properties of the specific functional architecture of the system. In most modern architectures of conventional computers, symbolic expressions are stored and later retrieved by their numeric or symbolic addresses (Pylyshyn, 1989, pp. 72–73). Although the set of computable functions does not depend on the particular implementation of the memory in the computing system, the time complexity of the computation executed does vary accordingly. Still, retrieving a particular string from a table can be, under certain conditions, be made

---

[13] This characterisation was at the heart of the classicist critique that whichever representations connectionist networks process – they do not exhibit compositionality and systematicity. By contrast, the operation of a *classical* computing system, according to the classicist, respects the semantic properties of symbolic expressions processed. This processing is akin to the processing of language-like inner representations in the brain that Fodor dubbed the "language of thought" (1975). We return to the tension between classicism and connectionism in Chapter 8.

independent of the number of strings stored and the size of the table (Pylyshyn, 1984, pp. 97–99).[14]

Another requirement, which may be dubbed the "maximal plasticity of function", seems to be alluded to by Pylyshyn in characterising concrete digital computing systems early on, but not in his later work (cf. Fodor & Pylyshyn, 1988, pp. 60–61). According to this requirement, the system has to *be programmable to allow the maximal plasticity of function*. Despite the rigidity of the physical structure of digital computing systems and the interconnections of their components, these systems exhibit maximal plasticity of function. This plasticity of function (or universality) is entailed by symbolic systems that may be programmed to behave in accordance with any finitely specifiable function (Pylyshyn, 1984, p. 53). This plasticity of function implies that a formal symbol manipulation system can produce any (Turing-computable) input-output function (Pylyshyn, 1984, p. 51).

Yet, it is not clear whether Pylyshyn reserves the term 'computer' just for those systems with maximal plasticity of function. For instance, Pylyshyn says that what he calls a 'computer' is a physical device "viewed as a variant of a Turing machine" (1984, pp. 70–71). In a later well-known critique of connectionism, Fodor and Pylyshyn summarise the basic point regarding classic digital computing systems as follows.

> "[N]ot all the functions of a Classical computer can be encoded in the form of an explicit program; some of them must be wired in. In fact, the entire program can be hard-wired in cases where it does not need to modify or otherwise examine itself. In such cases, Classical machines can be *rule implicit* with respect to their programs" (Fodor & Pylyshyn, 1988, p. 60, italics original).

As observed in Chapter 3, a special-purpose TM is hardwired for a specific task by modifying the machine head's internal wiring. A special-purpose TM fits the description above in that its entire "program" can be hardwired, for it need not modify itself. Its rules are, therefore, *implicit*.

Fodor and Haugeland certainly do not restrict the class of digital computing systems to those that are *fully programmable*. Fodor, for one, asserts that any complex system changing its physical state in accordance with the laws of physics *is* a computer, as long as there exists some mapping from physical states of the system to "formulae in a computing language" that preserves the desired semantic relation among the formulae (1975, p. 73). Such computer *can*, but *need not*, be a universal machine. Similarly, according to Haugeland, whilst the computing systems that are of importance to the design of cognition *are* universal, they *need not* be. He emphasised that there is an important difference between a "special

---

[14] Consider, for example, a well-known data structure in computer science: the hash table. Hash tables use a hash function to map identifying keys (such as an employee's name) to their linked values (e.g., the employee's position in the company). In a well-sized table with the right hash function, the average efficiency of each value lookup is *independent* of the number of key-value mappings stored in the table.

[purpose] computer [built] out of hardware" and a "general purpose computer [...]
that will [...] act exactly as if it were the special computer" (Haugeland, 1997, p.
12). Haugeland also referred to the TM broadly, not *just* the UTM, as a "kind of
computer – a new basic architecture" (Haugeland, 1981, p. 133).

On a final note, Pylyshyn claims that another requirement has to be included to
specify what makes it the case that a symbol S represents, say, a particular daisy,
rather than something else (personal communication). CTM have typically been
missing that part. Specifically, he argues that the minimum function needed for
this representation relation to obtain is that some causal or nomologically
supported dependency holds between the daisy and S (Pylyshyn, 2007, p. 82). In
other words, according to this claim, another requirement needs to be included for
specifying how extrinsic representation obtains in computing systems.

5.2.3   The FSM Account Evaluated

Before turning to evaluate the adequacy of the FSM account, let us briefly pause
to discuss some similarities between the FSM and PSS accounts. Both these
accounts are grounded in *symbolic* computation and underscore the important role
external *semantics* plays in digital computation. The first requirement of the FSM
account emphasises that symbolic expressions are manipulated according to
formal rules and need to remain *semantically interpretable* even following
numerous transformation operations. The fifth requirement of the PSS account
emphasises that symbolic expressions are manipulated *in virtue of* their *semantics*.
These two accounts diverge on how semantics enters the computational process.
According to the FSM account, symbolic expressions are formally manipulated *in
virtue of* their *syntax*, but they are always semantically interpretable, since the
transformation rules are supposedly "truth-preserving". The semantics of the
manipulated symbolic expressions is epiphenomenal on their syntax.

On the other hand, according to the PSS account, computational processes are
causally affected by the *semantics* of the manipulable symbolic expressions. The
operation of a computational process may depend on some possibly distal entity,
which is designated by some input to the process. Yet, how this designation
relation is obtained between a symbol and some distal entity remains vague. As
observed above, Newell and Simon *might* have taken it for granted that symbols
symbolise by definition. If that is indeed the case, a deep metaphysical enquiry is
in order as Smith has been arguing. In the absence of a definitive answer to this
question, two options remain for accounts of concrete computation as an
intentional phenomenon. The first option is to accept such accounts as possibly
adequate, subject to satisfying other criteria, with the question of designation
remaining open, awaiting an answer to the metaphysical enquiry. The second
option is to reject them as inadequate and give up the possibility of finding an
adequate account of computation as an intentional phenomenon. Smith opts for
the second one. The approach advocated in this book is to reject the semantic view
of computation as shall become more apparent in Section 5.4.

Another seemingly important similarity between the FSM and PSS accounts is that digital computing systems engage in information processing at the symbol level. Fodor and Pylyshyn, for instance, claim that "conventional computers typically operate in a 'linguistic mode', inasmuch as they process information by operating on syntactically structured expressions" (1988, p. 52). Newell and Simon's fundamental working assumption was that "the programmed computer and human problem solver are both species belonging to the genus IPS [information processing systems]" (1972, p. 870). The next chapter is devoted to exploring the manner in which digital computation is explicable in terms of information processing. It seems clear right from the start though, that both the FSM and PSS accounts rely on a semantic, rather than a quantitative, conception of information that is not limited only to intrinsic or mathematical representation.

Let us now turn to evaluate the FSM account based on the adequacy criteria. In a similar vein to the PSS account, it is too narrow, because it excludes too many paradigmatic digital computing systems. Therefore, the FSM account likewise fails to meet the dichotomy criterion. The focus of this account is on symbolic computation in which explicitness is inherently linked with symbols. According to Fodor and Pylyshyn, discrete connectionist networks lack the capacity to employ a compositional syntax for any representations that are processed and the capacity to preserve the semantics of the constituents in compound structures (1988). By their lights, such networks do not inherently invoke explicit representations, unlike classical symbolic computing systems. Accordingly, discrete connectionist networks fail to meet the first and fourth requirements of the FSM account.

The FSM account also excludes logic gates, logic circuits and FSA from the class of digital computing systems proper. The following excerpt from Pylyshyn (1984) on the distinction between an FSA and what he calls a 'computer' reveals the reason for that exclusion.

> "The difference between an extremely complex device characterized merely as proceeding through distinguishable states (but not processing symbols) and what I call a 'computer' is precisely the difference between a device viewed as a complex finite-state automaton and one viewed as a variant of a Turing machine. I shall use the term *finite-state automaton* as a general way of talking about any device whose operation is described without reference to the *application of rules to symbolic expressions*" (Pylyshyn, 1984, pp. 70–71, italics added).

On this view, unlike a "symbolic computer", an FSA does not perform any operation on symbolic expressions. Instead, the behaviour of the FSA can be fully explained by appealing to its state-transition regularities. The key difference in this context is that the function computed by a TM is supposedly *semantically interpretable*, but the one computed by an FSA is not. According to Pylyshyn, since a TM has a tape, which may be extended without bounds, it can be described in semantic terms, such as computing some numerical function or playing chess. An FSA, on the other hand, might have an arbitrarily large number of states, yet it cannot be similarly described in semantic terms. For such an FSA will not be amenable to a "finitary characterization of [its] state-transition regularities [and thereby...] not allow us to understand [...] the function such a machine is computing" (Pylyshyn, 1984, p. 72). An FSA violates the fifth requirement of the FSM account by not having a capacity to *write* to memory.

The resulting class of computing systems denoted by the FSM account is restricted to either symbolic *programmable stored-program* computers or symbolic *program-controlled* computing systems. If the resulting class is the former, the FSM account and the PSS account denote the same class of computing systems. On the other hand, if the FSM account does not mandate the universality requirement, it also includes special-purpose computers and (physical instantiations of) TMs. Even then, it remains *too* narrow. It excludes FSA, logic gates, flip-flops, shift registers and discrete connectionist networks, for they are not *symbolic program-controlled* systems.

As the result of being too narrow the FSM account, at best, only partially meets the taxonomy criterion. If it only classifies symbolic program-controlled systems as genuinely computational, it can only use the explicitness versus implicitness of rules (or "programs") as a differentiator among computing systems. On the FSM account, stored-program computers (i.e., physical approximations of UTMs) follow *explicit* rules. On the other hand, program-controlled computers, which are not stored-program (i.e., physical instantiations of special-purpose TMs), follow *implicit* rules. But this taxonomy is very limited, as it leaves many other computing systems unaccounted for.

Whilst the FSM account is sufficiently conceptually rich to explain miscomputations, it does not explicitly account for miscomputations. Pylyshyn attributes some errors, such as faulty pixels in the computer's display screen, to the "physical level" (1999, p. 9).[15] An explanation of such errors has to be given "in terms that refer to electrical and mechanical things — they require physical level explanations" (ibid). Haugeland wrote that "[i]t is often [wrongfully] said that computers never make mistakes (unless there is a bug in some program or a hardware malfunction)" (1997, p. 14). The question that should be asked is what counts as a mistake, that is, relative to which result specification (Haugeland, 1981, p. 18). Accordingly, by his lights, a chess-playing computing system can function without making a single error in following its internal *heuristics*. Yet, the system may be making many errors in playing the game, because its heuristics are clumsy. However, Haugeland also argued that in some sense "every digital

---

[15] Pylyshyn identifies a tripartite hierarchy concerning the organisation of digital computing systems (1984, pp. 93–101, 1989, pp. 58–59). The top level is the semantic, or knowledge, level, in which computational structures and processes are described in terms of their representational contents. The middle level is the symbol, or algorithm, level, in which computational structures and processes are described in terms of their syntax without any regard to their representational contents (i.e., purely formally). The bottom level is the physical level, or the functional architecture of the system, in which the physical substrate and the set of supported primitives are specified and underlie the structures and processes at the middle level.

computation (that does not consult a randomizer) is algorithmic; so how can [... it also] be heuristic? The answer is [... that] whether any given procedure is algorithmic or heuristic depends on how you describe the task" (1997, p. 14). This distinction between heuristics and algorithms makes the explanation of errors even less obvious.

Furthermore, what seems curious is how errors at the symbol level would be explained by the FSM account. Mathematicians and computer scientists attempt "to prove mathematically that a certain program will always generate outputs that are consistent with a certain mathematically defined function" (Pylyshyn, 1999, p. 9). And for this purpose, so Pylyshyn argues, the computing system has to be described semantically in terms of the things it represents. According to the FSM account, computational processes are characterised as rule-governed transformations of *interpretable symbolic expressions* (as per the first requirement above). It seems plausible then that *miscomputation* may be identified with *misrepresentation*, on this account.

However, identifying miscomputation with misrepresentation seems wrong. By way of explanation, let us consider a simple example. Suppose that a computing system CS performs computation C_1 at time T_1 on input I_1 producing O_1 and then performs some other computation C_2 at T_2 on O_1 producing some output O_2 at T_3. Let us further assume that a physical malfunction occurs that, nevertheless, happens to produce output O_2 at T_3. CS performs C_1 at T_1 as before, but it malfunctions and then performs C_3, instead of C_2, at T_2 on O_1 while still producing the same output O_2 at T_3. This is analogous to a two-step deductive argument $(P{\rightarrow}R{\rightarrow}Q)$, where the first step $(P{\rightarrow}R)$ is valid. The second step $(R{\rightarrow}Q)$ is *invalid* and yet the overall argument $(P{\rightarrow}Q)$ is *valid*. Although the final outcome is the same as the intended one (when it functions properly), we would still say that CS miscomputed.

How would a proponent of the FSM account describe this miscomputation? She can reply that either O_2 still respects the intended semantic interpretation, despite the malfunction, or O_2 is indeed a misrepresentation in this case. But had CS not malfunctioned, CS would have produced O_2 correctly, so, by her lights, O_2 respects its intended semantic interpretation. Since the same O_2 is produced as output when the system malfunctions, O_2 cannot qualify as a misrepresentation. And if she acknowledges that O_2 is not a misrepresentation, despite the system's malfunction, then misrepresentation and miscomputation cannot be identified. CS was simply subject to unpredictable noise that fortuitously resulted in the same output O_2. In the absence of a clear indication of how miscomputations are to be explained on the FSM account, we may conclude that it fails to meet the miscomputation criterion.

On the other hand, the FSM account seems to meet both the conceptual criterion and program execution criterion. As for the former, notions such as 'algorithm', 'program', 'architecture' and 'data' play a key role in the FSM account, as we have seen above. As regards the latter, as evidenced by the very first requirement of the FSM account, physical systems are computational in virtue of executing programs. Those programs are identified with internally represented rules, which govern the behaviour of the computing system. On the FSM account, digital computation is equivalent to program execution.

In sum, whilst the FSM account is ultimately found inadequate for explaining concrete digital computation, it offers many insights about those digital computing systems that execute programs. This account is indeed on the right track in viewing digital computing systems as automated logic-based systems. Many theoretical conceptions at the heart of FSM are fundamental to computer science. Importantly, the FSM account is also the foundation of classical computationalism, which was the reigning paradigm in cognitive science for several decades. However, besides being too narrow, this account extends beyond the mere formal operations of programs on uninterpreted data into the realm of external semantics. The processing of *meaningless* symbols is one thing. But insisting on the processing of *semantically interpreted* symbols in a manner that is truth preserving, at the semantic level, is another thing altogether.

5.3 A Reconstruction of Smith's Participatory Account

5.3.1 Introduction

The very first thing to note regarding this account is that it is by no means explicitly advocated by Smith as such (cf. fn. 2 above). For advocating this account goes against his fundamental negative claim that there *is no* and *never will* be a satisfactory and intellectually productive account of concrete digital computation (Smith, 2002, p. 24, 2010, p. 38). Nevertheless, in this section we examine an outline of a possible account of computation that *could* be reconstructed from Smith's positive claims about computation as a participatory phenomenon. These claims are essentially the basis of his criticism of the extant accounts of computation, such as effective computability (i.e., Turing's analysis examined in Chapter 3), the FSM and PSS accounts presented in the preceding sections, the algorithm execution account (a version of which is presented in Chapter 7) and the digital state machine account. With this caveat in mind, for simplicity we shall simply refer to this account as 'Smith's participatory account'.

Digital computers, according to Smith, are concrete, situated entities that participate and can do real work in the material world (2002, p. 37, 2010, p. 17). Computing systems, including those, which underlie modern real-time embedded operating systems in packet routers (that relay Internet traffic around the world) and aircraft navigation systems, are as *concrete* and *situated* as it gets (Smith, 2010, p. 9). In accordance with the empirical criterion examined in Chapter 2, Smith argues that real-world computing systems are extended to include a variety of peripheral modules. These peripheral modules, such as keyboards, screens, printers, network cables and Bluetooth operated devices, are controlled by the computer's operating system. These modules function as transducers through which the computing system actively participates in and interacts with the world that indiscriminately includes it (Smith, 2002, p. 37).

Furthermore, transducers enable the interaction of the computing system with its environment. Proponents of the FSM account may advocate two possible types of transducer: a physical one that crosses the physical boundaries of the system and a semantic one that crosses the semantic boundaries between symbols and their referents. By contrast, according to Smith, the only plausible type in practice is the physical type (2002, p. 39). However, by his lights, the resulting model of computation on the FSM account is of a "theorem prover plus [a] transducer" (Smith, 1996, p. 70). But the plausibility of this model depends on a rarely articulated assumption, namely, that the physical and semantic boundaries align, with symbols being inside the computing system and their referents outside. Yet, this alleged alignment does not occur in real-world computing systems in which the boundaries "*cross-cut* in myriad complex ways" (ibid, italics original).

According to Smith, this "cross-cutting of boundaries" essentially characterises any participatory system in general (1996, p. 70). This characteristic entails that concrete computation is functionally defined in terms of the participatory role it plays in the overall setting in which the computing system is situated. Besides, to *define* a semantic transducer, and not deem it genuinely computational, would make 'computation' simply shorthand for antisemantical symbol manipulation (Smith, 2002, pp. 39–40). Still, the manner in which 'transducer' is typically invoked in engineering signifies a mechanism that mediates changes in *medium* (e.g., an analogue-to-digital signal conversion), and not necessarily one that crosses the physical boundaries of the computing system (Smith, 2002, p. 57, fn. 28). On Smith's view, the traditional physical boundaries of a real-world digital computer do not block it from interacting with the environment in which it is situated.

5.3.2   The Key Requirements Implied by Smith's Participatory Account

Smith's participatory account identifies three key requirements for a physical system to perform digital computation. As remarked above, whilst these requirements may be taken as necessary for concrete computation, they cannot be sufficient.

- 1ˢᵗ Requirement: being situated in some environment.
- 2ⁿᵈ Requirement: standing in semantic relations to distal states of affairs.
- 3ʳᵈ Requirement: the processes of the system extending beyond its physical boundaries.

The first requirement, that the system be situated in some environment, seems trivially true at first glance. For this requirement similarly holds for *any* physical object whatsoever. Let us further elaborate what it amounts to in the context of computing systems. On Smith's view, computing systems are essentially situated in their environment, rather than just *operating in isolation*, where their situatedness is merely contingent. Although the behaviour of computing systems is effective, it is not limited to the physical boundaries of the system. According to Smith, computer scientists and practitioners do not simply deal with abstract notions of *computability*, they also deal with metaphysical questions about the ontology of the world in which computing systems are embedded (1996, p. 42). One obvious reason for that is the need to understand how the functioning system affects and is affected by the environment in which it is embedded (i.e., the "context" in which the system is deployed). Another less obvious reason is that in being *intentional* a computing system will *represent* at least some aspects of this context. That is, computing systems also process extrinsic representations essentially.

Moreover, on Smith's view, physical digital computing systems deal with issues of embodiment, situated interaction, physical implementation and, as the next requirement reveals, external semantics. The logical notion of computability simply does not address the situatedness aspects of physical computing systems. Real-world computing systems, on his view, cut across both physical and *semantic* boundaries. "It is not just that computers are involved in an engaged, participatory way with *external* subject matters, [... they] are participatorily engaged in the world *as a whole* – in a world that indiscriminately includes themselves" (Smith, 2002, p. 37). The result then is that computing systems actively interact with their environment, rather than just operate in a purely "internal world" of symbol manipulation. In this respect, Smith's view resembles Newell and Simon's view of physical symbol systems being *situated* in the physical world.

Smith's view of the essential situatedness of computing systems also bears resemblance to that of Andrew Wells'. The latter argues that (physical instantiations of) TMs, and likewise computers, are *embedded* in an environment that constitutes part of the TM's architecture (Wells, 1998, p. 280). Wells advocates an *externalist* rather than an *internalist* interpretation of Turing's account (1998, pp. 270–271). On his view, whilst the finite state controller is part of the TM's internal architecture, the TM's tape should be deemed a part of the *external environment*. TMs should, thus, be viewed as being irreducibly "world involving" (Wells, 1998, p. 280). Real-world computing systems are situated in and interact with their surrounding environment. Wells argues that Turing indeed thought of the machine's tape as instantiated in a medium that is different from the controller. This is because human memory is limited by its supervenience on a *finite* substrate, whereas the TM's tape is *infinite* (Wells, 1998, p. 282).

The second requirement is the system standing in semantic relations to *distal* states of affairs. According to Smith, this requirement reflects the representational character of computing systems. Although it is typically assumed that computation merely operates effectively on symbols, on his view, this assumption is wrong (Smith, 1996, pp. 70–71, 2002, pp. 34–35, 2010, pp. 28–29). Smith argues that computing systems stand in semantic relations to distal and other non-effective states of affairs. In other words, this requirement suggests that digital computing systems are not confined to operating on just *intrinsic* and *mathematical* representations, but they also operate on *extrinsic* representations.

As remarked above, by Smith's lights, computing systems cut across both physical and semantic boundaries. Consequently, manipulated symbols do not just affect *internal* referents (for instance, a memory address and the instruction in the memory to be executed next), but also *external* referents (for example, a numeral stored in the payroll database and a corresponding numeric value of an employee's salary). The crossing of the semantic boundary makes the computing system representational, unlike, say, a wall. When a symbolic expression is manipulated by the computing system causing, for example, a numeric value in the payroll database to change, all going well, this change eventually leads to the employee getting a pay rise, or less fortunately a pay cut, in the real world.

Moreover, the efficacy that constrains concrete computation has direct consequences in the real world. Smith argues that situated computing systems are consequential players in the very world they represent (2010, pp. 23, 29). In their operation, they make effective use of and affect the very states of affairs that their symbols extrinsically represent. Even the simple case of a loop counter that returns as output the number of unique type elements that were entered as input demonstrates, according to Smith, the crossing of the semantic boundary (1996, p. 71). For example, the numeral '3' would be returned as output, when given as input <'dog', 'cat' 'mouse', 'dog'> following a series of effectively related computational steps of the loop counter. A long downtime of a backbone mail-server, and an auto-navigated vehicle losing its set course, exemplify how transitions between local internal states of computing systems have long-distance causal affects on some states of affairs (Smith, 2002, p. 37, 2010, p. 36).

The third requirement is the processes of the system extending beyond its physical boundaries. Smith asserts that whilst the *internal* states of a computing system operate effectively (i.e., they are causally influential), computational processes typically extend beyond the physical boundaries of the system (2010, pp. 29, 36). Whether one examines computation as a process, which may even take place in an abstract TM, or considers its physical actualisation in a concrete system, an efficacy relation holds among the internal states of the system. This is how a computational process moves from its initial state and possibly input, through state-transitions to possibly producing output and terminating at the final state.

Moreover, the computing system involves causal interactions among internal symbols and referents, which can be either internal (e.g., a memory address) or external (e.g., a speeding car caught on camera) (Smith, 2010, p. 29). Even if we imposed theoretical boundaries between the system and its surrounding environment, these, by Smith's lights, would not be boundaries to the flow of effect. A traffic monitoring computing system, for instance, is affected by and affects the traffic on the road as an external referent. Likewise, a signal, which is sent by one computing process across to another external system, "travels" beyond the physical boundaries of the source system and awaits an acknowledgement response from the target system. The flow of effect continues despite the physical boundaries of the two systems.

On Smith's view, real-world computing systems need to be analysed in a middle ground realm between the abstract and the concrete (1996, p. 36). The analysis required is of a metaphysical nature and has not been recognised yet by the natural sciences. Computing systems need to be understood at a level more abstract than is typical in classical physics, but also more concrete than is typical in classical computability theory. The physical boundaries of the computing system divide what is *internal* to the computing system (e.g., control unit, memory and input/output devices) from what is *external* (i.e., the environment in which the system is embedded). The semantic boundaries divide the realm of *symbols* being manipulated during a computation from their *referents,* which can be either internal (e.g., instructions stored in the system's memory) or external (e.g., employees' salaries) (Smith, 1996, pp. 70–71, 2002, pp. 35–36).

Yet, physical computing processes are not limited by the semantic and physical boundaries to symbol manipulation only occurring in some internal world of symbols and referents. They are *participatory,* despite any theoretical boundaries that separate such processes from an external realm of referents (Smith, 2002, p. 40, 2010, p. 29). In this sense, according to Smith, they extend beyond the boundaries that apply to formal systems, such as axiom systems about numbers, and involve complex paths of causal interaction among symbols and referents (both internal and external), which are cross-coupled in complex configurations. Smith emphasises that this does not imply that *semantical properties* of computing systems are *causally efficacious.* Rather, this crossing of boundaries shows that the participatory involvement of computing processes is sufficiently strong to resist any attempts to define them as being merely *formal* (Smith, 2010, p. 29).

5.3.3 Smith's Participatory Account Evaluated

On Smith's view, Turing's analysis is reduced to "a mathematical theory of causality" (2002, p. 42, 2010, p. 23) and fails as *an account of computation.*[16] For computation is both constrained by and constituted by physical relations between the computing system and the environment. Turing's analysis provides a mathematical theory of the physical world: how computationally hard it is for one physical configuration to change into another by means of either scanning and writing symbols or some other finite physical processes (Smith, 2010, p. 27). Instead, on a reconstructive reading of Smith, a concrete computing system satisfies at least the three requirements above and this implies that the system is *effective* but also *representational.* Where does that leave us in terms of evaluating the adequacy of the participatory account then?

---

[16] The reader should note that here Smith refers to Turing's original analysis and not to Turing's account presented in Chapter 3.

The three requirements above are focused on computing systems being representational entities situated in a physical environment, but they do not suffice for individuating genuine computing systems *qua* computational. In other words, these requirements can only be considered as an extension of some more "basic" account of digital computation. However, it is hard to determine what such a "basic" account would be. For Smith explicitly rejects all the extant accounts of computation and the possibility of providing *any* adequate account in principle. Yet, it seems clear that if we examined these three requirements in isolation, digital systems, such as thermometers, DVD players and CRT TV sets, would qualify as genuine computing systems. A digital thermometer, for instance, is situated in the environment when fulfilling its purpose (or else it would be useless). The process of measuring the temperature of some physical object, strictly, extends beyond the physical boundaries of the thermometer. It also stands in semantic relations to some distal states of affairs in that it may indicate, say, that some patient is running a high fever.

One plausible "basic" account of computation, which may be extended by the three requirements above, is an *externalist*, rather than an internalist, interpretation of Turing's account. An externalist interpretation of Turing's account, as Wells suggests, considers the TM's finite state controller as part of the TM's internal architecture, whilst it considers the tape a part of the external environment (1998, pp. 270–271). Further, to be compatible with Smith's view of computation as an interactive process, Turing's account would have to include the "oracle" component. The requirements of the participatory account would then supplement the five requirements of the Turing account examined in Chapter 3. Such a synthesis would imply unambivalently that computing systems are intentional and participatory, rather than just systems that arrange, move and transform "concrete physical stuff [...] by finite physical processes" (Smith, 2010, p. 27).

Another plausible "basic" account of computation, which may be extended by the three requirements of the participatory account, is that of *symbol manipulation* (rather than FSM). Smith argues that "the formal symbol manipulation construal [... fails] not [....] because it fails to be about computing, which on the contrary it genuinely is, but rather because it is *false*" (2010, p. 28, italics original). The FSM account ultimately fails, because concrete digital computation "turns out *not to be formal*" (ibid, italics original). However, it is not a simple task to exclude the *formality* ingredient from the FSM account, say, by excluding specific requirements of that account, in order to supplement the FSM account with the requirements of the participatory account.

So to evaluate the participatory account, we shall adopt the first option, thereby grounding it in an externalist interpretation of Turing's account. Only an evaluation based on the particular requirements, which the participatory account brings to the table, seems legitimate. Therefore, the present strategy is to evaluate whether any of the three supplementary requirements (or a combination thereof) leads to the synthesised *Turing participatory account*[17] not failing to meet a criterion, which the Turing account in isolation fails to meet.

The Turing account fails to meet the implementation criterion, and none of the three requirements of Smith's participatory account changes that. The first requirement of Smith's participatory account clearly implies that the computing system *is* physical. The second and third requirements combined imply that computational processes extend beyond the physical boundaries of the system and (somehow) affect distal states of affairs. Yet, neither of these two requirements specifies how abstract computation is actualised in the physical substrate.

In like manner, the Turing account fails to meet the miscomputation criterion, and none of the three requirements of Smith's participatory account changes that. In emphasising the concreteness of physical computing systems, the participatory account is certainly a step in the right direction. Disconnecting the system from the environment, in which it is situated, may result in a miscomputation in some cases, but not always. For instance, consider a computing system that performs some computation that relies on input from the environment via a transducer that is connected to the system. Either disconnecting the transducer from the system or somehow disconnecting the system from its environment (i.e., disallowing the normal operation of the transducer) will ultimately result in a miscomputation. Yet, a stand-alone computer, not hooked up to anything, such as the Altair 8800[18], need not miscompute simply because it is not "situated in the environment" (as long as it has a constant supply of energy).

Another possibility is that a computing system miscomputes, according to the participatory account, when the system *misrepresents* some distal state of affairs. That is, if a system fails to meet the second requirement of the participatory account, it miscomputes. This is analogous to the fifth requirement of the PSS account and the semantic interpretability of manipulated symbols as per the first requirement of the FSM account. However, as already observed in the critique of the FSM account, a plain identification of miscomputation with misrepresentation fails.

Where the participatory account fares poorly, contrary to expectations, is in meeting the dichotomy criterion. To reprise, the Turing account fails to meet this criterion when 'symbols' and 'instruction-following' are construed broadly enough. An externalist interpretation of Turing's account is unlikely to remedy this failure. By using the three requirements of the participatory account exclusively for distinguishing paradigmatic computing systems from paradigmatic

---

[17] The overall Turing participatory account is a synthesis of the Turing account (as examined in Chapter 3) and the supplementary characteristics of computation that Smith brings to the fore. Put another way, the Turing participatory account includes *eight* requirements overall.

[18] The Altair 8800 was a microcomputer design from 1975 based on the Intel 8080 CPU. The Altair 8800 computer did not have any installed software. The operator used switches on the computer's front panel to enter instruction codes and data (Allan, 2001, pp. 6–5ff).

non-computing systems, the resulting class of genuine digital computing systems is *too* broad. For the participatory account classifies keyboards, screens, printers, digital thermometers, DVD players, modems and remote controls as genuinely computational.

Let us briefly examine why such systems qualify as computational by meeting the three requirements of the participatory account. We have already seen why a digital thermometer meets those requirements. A keyboard is likewise situated in the environment when fulfilling its purpose (specifically, it is connected to a conventional computer). When its keys are pressed, provided that it is connected to an operating computer, discrete signals are sent to the computer thereby extending beyond the physical boundaries of the keyboard. It also stands in a semantic relation to some state of affairs that is determined by the particular program that the computer is running (e.g., the pressing of a button that sends the *restart* command to the computer's OS). The same can be easily shown for screens, printers, DVD players, modems and remote controls. From this we may extrapolate that too many physical systems are classified as genuine digital computing systems, thereby showing that the participatory account, as such, fails to meet the dichotomy criterion.

The participatory account does no better in meeting the taxonomy criterion. Whereas the Turing account can, at the very least, taxonomise computing systems relative to whether or not they are universal, the participatory account offers no clear way of taxonomising them. The three requirements of the participatory account apply, according to Smith, equally well to any computing system. A computing system cannot be more or less situated in the environment. Similarly, the system either stands in semantic relations to some distal state of affairs or not. But neither of these requirements serves as an indication of the computing powers of computers.

In sum, perhaps none of the criticisms of the participatory account above would come as a surprise to Smith, given that he argues that "[w]e will never have a theory of computing" (2010, p. 39). The goal of his ambitious project has been to find the right characterisation that does justice to computational practice, the theoretical foundations of computability and CTM simultaneously. To achieve this goal, he has subjected the extant construals of computation to severe scrutiny, ultimately judging them all inadequate as *accounts* of computation. Yet, despite this seemingly negative result, Smith teaches us an important lesson in how quickly we employ the notion of computation without knowing precisely what we mean by that or what computation is.

5.4 Arguments for the Semantic View of Computation Criticised

CTM are based by and large on the premise that cognition and computers have a great deal in common. Proponents of the semantic view of computation seem to consider external semantics as the necessary platform that can bridge the gap between a digital computer and cognition. This premise was certainly at the core of the PSS and FSM accounts of digital computation. It also inspired Smith to insist on the cognitive criterion for judging the adequacy of accounts of computation, as we have seen in Chapter 2. By way of concluding this chapter, we shall now turn to evaluate some of the main arguments in support of the semantic view of computation. Three such arguments are presented and criticised.

The ensuing discussion deals mostly with intrinsic and extrinsic representations, yet this distinction can be made more granular. It can be adjusted, for example, to cater for intercommunication among digital computing systems such as in distributed networks of computers and among computing systems that are connected to transducers that allow a representation of their environments, such as in robots. Such a representation is arguably no more problematic than intrinsic representations in computing systems. For the representation of either the state of an interconnected computing system or a physical object in the environment, captured through a sensor attached to the robot, is the result of some causal relations with the environment. However, the discussion of this fine-grained type of representation exceeds the scope of this book.

The next three arguments, in one form or another, purportedly justify the *semantic individuation* of digital computation or computational states. To claim that digital computational states are semantically individuated is the same as claiming that these states are individuated or classified by their semantic, or representational, properties. Put yet another way, this is tantamount to claiming that "computational states have their content essentially" (Piccinini, 2008, p. 205).

The first argument, which may be dubbed the "semantic individuation of mental states", uses two basic premises.

- (P1) Computational cognitive science resorts to digital computational states in trying to explain the nature of mental states.
- (P2) Mental states are semantically individuated.
- Therefore, (C) for the purposes of adequate explanation, digital computational states should be semantically individuated (Burge, 1986, pp. 28–29; Peacocke, 1999, pp. 197–198; Pylyshyn, 1984, pp. 57–58, 1989, pp. 55–57)

This argument is invalid. Either a third premise has to be assumed (Egan, 1995, p. 183; Piccinini, 2008, pp. 226–227) or one of its two premises P1 or P2 needs to be strengthened in some way, so that the conclusion would follow logically from the premises. A sufficient third premise could be that the explanans has to be individuated in the same way that the explanandum is. In order that a computational explanatory framework will be successful in explaining mental states, digital computational states have to be semantically individuated too. Alternatively, the first premise could be strengthened in such a way that it would imply that mental states are *equivalent* to digital computational states.

However, both these alternatives simply emphasise the weakness of this argument. Semantic individuation is only one possible dimension of individuation for both computational and mental states (other dimensions of individuation might be, for example, functional or physical). Proponents of this argument should further assume either the equivalence of mental states and computational states or, at the very least, that computational states have to be individuated in the same way that mental states are. But even then, the semantic individuation of computational states might still be very different from that of mental states. Bearing in mind that the original intent of the arguers was to gain considerable explanatory leverage by resorting to digital computation, the whole endeavour is weakened on pain of circularity. Another line of argument is needed to justify the conclusion.

The second argument, which may be dubbed "computational identity", uses three premises.

- (P1) Digital computation is individuated by the *functions* being computed.
- (P2) Functions are defined by *syntactic* structures generating output $F(I)$ for input I.
- (P3) The *underlying* function being computed *in a given context* is, at least partially, semantically individuated.[19]
- Therefore, (C) digital computation is, at least partially, semantically individuated (Dietrich, 1989, pp. 119, 126–128; Shagrir, 1999, pp. 137–143; Smith, 1996, pp. 6–9).[20]

There is, at least a prima facie, conflict between conclusion (C) and the *one that could be drawn* from the first premise (P1) and the second premise (P2). The first two premises suggest that digital computation is individuated syntactically. The problem is then which conclusion we should adopt. There is also an obvious tension between the first premise (P1) and the third premise (P3). If we chose to discard the third premise (P3), then the conclusion (C) that digital computation is, at least partially, semantically individuated would not follow. As a result, we get two different answers to the question how digital computation is individuated. One answer is that digital computation is individuated *syntactically*. The second answer is that digital computation is individuated *semantically*.

The tension between the first and the third premises can be supposedly resolved in the following way. A digital computing system may implement two or more functions simultaneously. This claim, combined with the second premise, entails that digital computation is individuated syntactically, for functions are *defined* syntactically. Let us grant that the computational identity of a digital computing system is indeed context dependent (Shagrir, 1999, p. 142). This dependency is on the *specific* function being computed, that is, the underlying function noted by the

---

[19] Put another way, the computational identity of a system is context-dependent: it depends on the syntactic structure the system implements when performing a given task (Shagrir, 1999, p. 142). The underlying function being computed in that context determines the computational identity of the system.

[20] It should be noted that Shagrir does not think that this is a valid argument, but that it indicates nevertheless (as evidenced by premise three) that the individuation of computation is, at least partially semantic (personal communication).

third premise. This context-dependency supposedly requires a semantic constraint, which can individuate digital computation relative to the given context. Since other syntactic structures may be extracted simultaneously, a constraint *external* to the computing system concerned is required to determine which of these syntactic structures defines the *computational identity* of the system (Shagrir, 1999, p. 143). This constraint, according to proponents of the "computational identity" argument, has something to do with the *contents* of the computation being executed.

Yet, this line of reasoning invites the question: why should the underlying function, which determines the computational identity of the system, be *semantically* individuated? It may indeed be *sufficient* to individuate digital computation semantically, but it is certainly *not necessary* (Egan, 1995, pp. 185–187; Piccinini, 2008, pp. 223–225). Shagrir, for one, argues that the functions a system computes are *always* characterised in semantic terms, viz., in terms of the contents of the system's computational states. So, for instance, "[w]e say that a system computes 34+56 (where the function '+' is defined over numbers, not numerals), or the next move of the white queen on the chess board" (Shagrir, 2001, pp. 382–383). The individuation of computational functions in terms of the *contents* of the system's states underscores the need to distinguish among types of computational representation as already observed in Chapter 1.

Extrinsic representation can be dispensed with, whereas intrinsic and mathematical representations cannot. It is extrinsic representation that proponents of the "computational identity" argument propose as an "external" constraint for determining the computational identity of the system. A system computing 34+56 makes use of a *mathematical* representation. The "move of the white queen on the chess board" can be explained it terms of data manipulation without invoking extrinsic representations (i.e., a white queen piece) necessarily. Consider again the two possible interpretations of a conventional AND gate discussed in Chapter 2. Even if we interpreted the operation of a particular logic gate as either a conjunction or disjunction operation (i.e., performed by an AND gate or an OR gate, respectively), the underlying data processing operation of the gate remains unchanged. It is only a matter of how we choose to *name* this operation, and consequently the logic gate, that varies semantically. Whilst designers and users of digital computers characterise the functions computed by the computer semantically, this characterisation is by *convention* rather than by *necessity*.

Besides, a semantic characterisation of computed functions *presupposes* a syntactic, or formal, characterisation. If that is right, then it is not clear whether one semantic characterisation has an epistemic privilege over another. As Piccinini points out, given a description of a computed function in terms of its input and output, the question arises of how it is that a computing system may compute a function, which is defined over numbers, strings, vectors or some other entities (2008, pp. 224–225). A characterisation in terms of input argument(s) and output denoting the domain and the range of the function, respectively, requires that the input and output be specified *formally*. In our terminology, such characterisation of computation is accomplished by invoking mathematical representations.

Moreover, any Turing-computable function may be implemented in a variety of different ways, thereby making semantic individuation problematic. The function computed by a physical system may be characterised in any number of ways relative to the corresponding level of abstraction. Viewed from a sufficiently high level of abstraction, a semantic individuation of a given program seems plausible. Nonetheless, even if the particular physical computing system is fixed, any Turing-computable function is computable by infinitely many algorithms. Each algorithm, in turn, is realisable by arbitrarily many programming languages. At the level of the machine code itself, any trace of the semantic description, given at the higher level of abstraction, is replaced with data transfers between registers (or some other technology-specific components). Still, the computational states of the system concerned can still be *formally* individuated. Accordingly, computational functions need not be semantically individuated.

The third argument, which may be dubbed the "multiplicity of computations", is a variation on the "computational identity" argument. The emphasis in the "computational identity" argument is on the *functions* computed, whereas in the "multiplicity of computations" argument the emphasis is on the computational *tasks*. Yet, a computational task is defined as computing a particular function, "for example, [... the] task of P is to compute addition" (Shagrir, 2001, p. 377). This subtle distinction does not seem to make any significant difference for the third argument. The "multiplicity of computations" argument is reasoning by implication, since Shagrir's original argument was specifically directed at cognitive systems broadly (2001, p. 374, fn. 6).

The "multiplicity of computations" argument can be summarised as follows.

- (P1) Any given digital computing system may simultaneously implement different computations.
- (P2) In any given context, the digital computation performed by a computing system is determined by a single syntactic structure, which is the underlying *task*.
- (P3) The underlying *task* is, at least partially, semantically individuated.
- Hence, (C) digital computation is, at least partially, semantically individuated relative to one particular task that the digital computing system performs.

Whilst, as observed above, the target of this argument is cognitive systems, Shagrir explicitly asserts that the first premise (P1) and the second premise (P2) pertain to digital computing systems in general (2001, pp. 376–377). He also points out that appealing to examples of non-cognitive systems reinforces this argument. By Shagrir's lights, digital computational states of the same physical system P can be classified as different syntactic structures when P is used for performing different computational tasks. "There must be another constraint [then], [that is] *external to* P [... determining] which syntactic structure constitutes the computational identity of P" (Shagrir, 2001, p. 384). Shagrir concludes that the only plausible constraint is provided by features of the derived content of P's computational states.

The first two premises are not problematic, but the third premise (P3) shoulders most of the burden and requires further evidence. It is certainly the case that a digital computing system may perform multiple computations, as a general-purpose computer clearly demonstrates (or a multitasking computer whose speed is measured in millions of instructions per second). P1 is self-explanatory. P2 rightly states that these multiple computations are determined by *syntactic structures*. But these two premises lead Shagrir to assert that some constraint needs to be employed to individuate digital computation relative to a specific task that the computing system performs (2001, p. 381). The argument starts by asserting that multiple computations are individuated *syntactically*, and proceeds by introducing a *semantic constraint* that supposedly has to be employed to determine which one of the computational tasks is performed in a given context.

This line of argument raises an important question: how is the explanatory computational context *defined*? As suggested by Piccinini, an underlying assumption of this argument is that the explanatory context is *extrinsic* (2008, p. 231). If the explanatory context is extrinsic, that is, external to the physical boundaries of the computing system, then an independent knower is required to interpret the computational states and output as extrinsic representations of objects or state of affairs. In this case, a particular task is interpreted semantically. Furthermore, when each individual computational task is individuated syntactically, then determining which of the multiple tasks is the computational identity of the computing system is open to interpretation. If every particular computational task can be individuated syntactically, then the semantic constraint may be *sufficient* to determine the computational identity of the system, but is *not necessary*.

Still, the given explanatory context can just as well be *intrinsic*, in which case a semantic constraint need not be employed. In this case, the particular computational task can also be individuated non-semantically. A conventional computer program, which is comprised of different subroutines, might respond differently to inputs, such as <'123', 'abc'> and <'a=1; b=2; c=3'>, and produce diverse outputs in response. However, the appropriate computational description of the program can be determined without employing any semantic constraints. This, of course, leaves open the question of how the computational identity of *cognitive systems*, being the *original* target of the third argument, is determined (provided that these systems are even computational to start with).

In sum, the arguments in support of the semantic view of computation fail to provide a compelling justification for individuating digital computation semantically. For one thing, it is a bad starting point to characterise computational states and structures in terms of representational content without a detailed explication of what this 'representation' even means (Ramsey, 2007, p. 8). This was already discussed in Chapter 1 by observing that representation is typically defined as a tripartite relation that holds among the representing entity, the represented entity and the interpretation (Peirce's interpretant). Newell and Simon defined representation in terms of a symbol's capacity to designate some distal entity. Yet, their definition of 'designation' does not explicate the relation that holds among the representing, or designating, entity, the represented, or designated, entity and the designation process. The FSM account and Smith's participatory account similarly characterise digital computation semantically. Nevertheless, despite the motivation to individuate computational states in a like manner as mental states, a semantic individuation of computational states is not well justified by the arguments examined in this chapter.

References

Allan, R.A.: A history of the personal computer the people and the technology. Allan Publishing, London (2001)

Bickhard, M.H., Terveen, L.: Foundational issues in artificial intelligence and cognitive science: impasse and solution. Elsevier, Amsterdam (1995)

Burge, T.: Individualism and Psychology. The Philosophical Review 95(1), 3–45 (1986), doi:10.2307/2185131

Clark, A.: Mindware: an introduction to the philosophy of cognitive science. Oxford University Press, New York (2001)

Copeland, B.J.: Computation. In: Floridi, L. (ed.) The Blackwell Guide to the Philosophy of Computing and Information, pp. 3–17. Blackwell, Malden (2004)

Dietrich, E.: Semantics and the computational paradigm in cognitive psychology. Synthese 79(1), 119–141 (1989), doi:10.1007/BF00873258

Egan, F.: Computation and Content. The Philosophical Review 104(2), 181–203 (1995), doi:10.2307/2185977

Fodor, J.A.: The language of thought. Harvard University Press, Cambridge (1975)

Fodor, J.A.: Methodological solipsism considered as a research strategy in cognitive psychology. Behavioral and Brain Sciences 3(01), 63–73 (1980), doi:10.1017/S0140525X00001771

Fodor, J.A., Pylyshyn, Z.W.: Connectionism and cognitive architecture: A critical analysis. Cognition 28(1-2), 3–71 (1988), doi:10.1016/0010-0277(88)90031-5

Haugeland, J.: Mind design: philosophy, psychology, artificial intelligence, 1st edn. MIT Press, Cambridge (1981)

Haugeland, J.: Artificial intelligence: the very idea. MIT Press, Cambridge (1985)

Haugeland, J.: Mind design II: philosophy, psychology, artificial intelligence (rev. and enl. ed.). MIT Press, Cambridge (1997)

Newell, A.: Physical Symbol Systems. Cognitive Science 4(2), 135–183 (1980a), doi:10.1207/s15516709cog0402_2

Newell, A.: The Knowledge Level. AI Magazine 2(2), 1–20 (1980b)

Newell, A., Simon, H.A.: Human problem solving. Prentice-Hall (1972)

Newell, A., Simon, H.A.: Computer science as empirical inquiry: symbols and search. Communications of the ACM 19(3), 113–126 (1976), doi:10.1145/360018.360022

Peacocke, C.: Computation as Involving Content: A Response to Egan. Mind and Language 14(2), 195–202 (1999), doi:10.1111/1468-0017.00109

Piccinini, G.: Computation without Representation. Philosophical Studies 137(2), 205–241 (2008), doi:10.1007/s11098-005-5385-4

Pylyshyn, Z.W.: Computation and cognition: toward a foundation for cognitive science. The MIT Press, Cambridge (1984)

Pylyshyn, Z.W.: Computing in Cognitive Science. In: Posner, M.I. (ed.) Foundations of Cognitive Science, pp. 49–92. MIT Press, Cambridge (1989)

Pylyshyn, Z.W.: What's in Your Mind? In: LePore, E., Pylyshyn, Z.W. (eds.) What is Cognitive Science?, pp. 1–25. Blackwell, Malden (1999)

Pylyshyn, Z.W.: Things and places: how the mind connects with the world. MIT Press, Cambridge (2007)

Ramsey, W.: Representation reconsidered. Cambridge University Press, Cambridge (2007)

Scheutz, M.: Computationalism – the next generation. In: Scheutz, M. (ed.) Computationalism: New Directions, pp. 1–22. MIT Press, Cambridge (2002)

Shagrir, O.: What is Computer Science About? The Monist 82(1), 131–149 (1999), doi:10.2307/27903625

Shagrir, O.: Content, computation and externalism. Mind 110(438), 369–400 (2001), doi:10.1093/mind/110.438.369

Smith, B.C.: Formal symbol manipulation: Ontological critique (forthcoming), http://www.ageofsignificance.org (retrieved)

Smith, B.C.: On the origin of objects. MIT Press, Cambridge (1996)

Smith, B.C.: The foundations of computing. In: Scheutz, M. (ed.) Computationalism: New Directions, pp. 23–58. MIT Press, Cambridge (2002)

Smith, B.C.: Age of significance: Introduction (2010), http://www.ageofsignificance.org (retrieved)

Wells, A.J.: Turing's Analysis of Computation and Theories of Cognitive Architecture. Cognitive Science 22(3), 269–294 (1998), doi:10.1207/s15516709cog2203_1

Chapter 6
Computation as Information Processing

6.1 Introduction

It seems incontrovertible that information processing is fundamental to cognitive function. As already observed in Chapter 1, one explanatory strategy for a theory of cognition to take is to view some of the agent's internal states and processes as carrying *information* about those relevant aspects of its body and external states of affairs in negotiating its environment (Bechtel, 1998, p. 297). Natural cognitive agents produce new information that is intended, amongst other things, to deal with a variety of environments. It is unsurprising then that semantic accounts of computation that underlie CTM view digital computing systems as engaging in information processing at the symbol level (cf. the FSM and PSS accounts).

It is now increasingly popular to view computers as "information processors". In the global information society, one of the key commodities that is traded on a worldwide scale is *information*. Information travels instantaneously across time and space through an array of new communication technologies and, particularly, the Internet. Computing systems make this all possible. This leads to the common assumption that computation can freely be described as information processing. The Encyclopedia of Computer Science states that "information processing might, not inaccurately, be defined as 'what computers do'" (Ralston, 2000, p. 856). This definition is put to the test in this chapter, dealing with the question whether concrete digital computation can be adequately explained solely in terms of information processing.

The resulting information processing (IP) account hinges on the particular conception of information that is used. Information can be construed semantically or non-semantically. Two of the most prominent non-semantic conceptions of information are Shannon information (introduced in the mathematical theory of communication) and algorithmic information (as it used in algorithmic information theory). Such conceptions deal with issues of signal transmission and signal compression, and in general with quantitative aspects of information. But information can also be construed in a manner closer to our intuitive usage of 'information' as representing objects or states of affairs, that is, as closely linked with *meaning* or *semantics*.

This chapter is organised as follows. In the next section, four possible conceptions of information are introduced: *Shannon* information, *algorithmic*

N. Fresco, *Physical Computation and Cognitive Science*,
Studies in Applied Philosophy, Epistemology and Rational Ethics 12,
DOI: 10.1007/978-3-642-41375-9_6, © Springer-Verlag Berlin Heidelberg 2014

information, and semantic content that may give rise to either *declarative* information or *instructional* information. Subsequently, in Section 6.3, features of the resulting IP accounts based on each particular conception of information are examined. Section 6.4 explicates the key requirements for a physical system to perform digital computation in terms of information processing. Section 6.5 examines the problems faced by the resulting IP accounts based on Shannon information, algorithmic information and declarative information (or more precisely, factual information). The main argument in Section 6.6 is that only an IP account based on instructional information is suitable for the task of explaining concrete computation. In the last section some observations are made regarding the systematic relations amongst the conceptions of information above.

6.2 Semantic Information, Non-semantic Information and Data

The roots of the conflation of information and computation are in the attempt to explain cognition in the mid-twentieth century. This venture led to a fusion of information-theoretic language, notions of control and TMs. At that time, the information-theoretic language used was based on the Shannon/Wiener cybernetic conception of information. The original ambition of cybernetics was to explain the behaviour of both living organisms and human-engineered machines using a unified theory utilising concepts of control and information (Cordeschi, 2004, p. 186). Norbert Wiener characterised a machine as "a device for converting incoming messages into outgoing messages [... or] as the engineer would say [...] a multiple-input multiple-output transducer" (Wiener, 1966, p. 32). Still, in general, the problems under consideration were based on single-input single-output transducers without any long-range statistical dependencies of transmitted messages.

However, our modern concept of information is broader than the cybernetic conception of information. To some degree, everything can be described in information-theoretic terms: from the movements of atoms and molecules through economics, politics and fine arts to ethics and human cognition. So broad is the concept of information that some construals of information unavoidably lead to pan-informationalism.

An important distinction that should be drawn in this context is between 'data' and 'information'. The Data – Information – Knowledge triad is a simple schema that has gained much attention in recent decades. The Encyclopedia of Computer Science defines *data* as "physical symbols used to represent the information" (Ralston, 2000, p. 502). *Information*, on the other hand, is typically defined as follows.

- "[T]he sum total of all the facts and ideas that are available to be known by somebody at a given moment in time" (Cleveland, 1982, p. 34).
- "[D]ata that have been interpreted and understood by the recipient of the message" (Lucey, 1987, p. 13).
- "[D]ata that has been given meaning by way of relational connection" (Ackoff, 1989, p. 3).

Data then are the vehicle that can convey meaningful information. In the following three sections, we examine four different data-centred conceptions of information.

In the present context, our focus on information is mainly confined to how information *processing* pertains to concrete digital computation. The processing of information is characterised here as encompassing the encoding, decoding, production, modification and removal of information. To a first approximation, such a characterisation seems to accord with the operations of a TM reading symbols from and writing symbols to a tape. Intuitively, this characterisation also seems to describe the practical use of information by natural cognitive agents beyond the mere communication of information. There are clearly other conceptions of information that may be relevant to our present enquiry, but we restrict our attention to four data-centred conceptions that are often used in computer science and in cognitive science.

6.2.1 Shannon Information as a Non-semantic Conception of Information

In the mathematical theory of communication (MTC), Claude Shannon attempted to solve the "fundamental problem of communication [...] that of reproducing at one point either exactly or approximately a message selected at another point" (1948, p. 379). According to MTC, one of the simplest unitary forms of information is the recording of a choice between some equiprobable basic alternatives (Shannon, 1948, p. 379; Wiener, 1948, p. 61). "The significant aspect is that the actual message is one *selected from a set* of possible messages" (Shannon, 1948, p. 379). These messages may consist of either sequences of letters (or symbols), as in telegraphy, for example, or, more generally, sets of signals. Shannon information (SI), then, is the simplest form of information as (statistically) structured signals (or as we call it below, structured data).

SI does not entail any semantic content or meaning. "[M]essages [may] have meaning; [but] [t]hese semantic aspects of communication are irrelevant to the engineering problem" (Shannon, 1948, p. 379). Shannon provided a statistical definition of information as well as some general theorems about the theoretical lower bounds of bit rates and the capacity of information flow. MTC approaches information quantitatively: how much information is conveyed. But SI tells us nothing about the usefulness of or interest in a message. The basic idea is coding messages for transmission at the bare minimum number of signals needed to get a message across in the presence of noise.

Even in this sense, the amount of information conveyed is as much a property of the sender and recipient's knowledge as anything in the transmitted message. If the same message is sent twice, that is, a message and its copy, the information in the two messages is not the sum of that in each. Rather the information only comes from the first one. Receiving a message may change the recipient's circumstance from not knowing what *something was* to knowing what *it is* (Feynman, 1996, pp. 118–120). According to MTC, information grows with the reduction of uncertainty it yields. When the recipient of some message is absolutely certain about a state of affairs, *E,* she cannot receive *new* information about *E* (Adriaans, 2012a). For this reason, SI links information and probability.

Lastly, SI is often compared with thermodynamic entropy. MTC quantifies statistical properties of the communication channel and the message selection process relative to a fixed set of messages. The more possible messages the set contains, the more uncertainty there is as to which message will be selected. Shannon entropy quantifies the uncertainty of the message selection process. It is a function that reduces a set of probabilities to a number, reflecting how many nonzero possibilities there are for selecting a message as well as the extent to which the set of nonzero probabilities is uniform. Within a fixed context, a smaller set of probabilities can be interpreted as more "orderly", in the sense that fewer numbers are required for specifying the set of possibilities. Thermodynamics dictates a particular context – probabilities have to be measured in the full state space of a given system. Thermodynamic entropy can be deemed a special case of Shannon entropy (Bais & Farmer, 2008, p. 636).

6.2.2 Algorithmic Information as a Non-semantic Conception of Information

Algorithmic information theory (AIT), which was developed by Andrei Kolmogorov, Ray Solomonoff and Gregory Chaitin, deals with the *informational* complexity of data structures (but not with the *computational* complexity of algorithms). It formally defines the *complexity,* that is, the *informational content* of a data structure, typically, a string, *x,* as the length of the shortest self-delimiting program[1] that produces *x* as output on a UTM (Chaitin, 2004, p. 107). The algorithmic information (AI) complexity of any computable string (in any particular symbolic representation) is the length of the shortest program that computes it on a particular UTM.

Moreover, Chaitin proposes thinking of a computing system as a decoding device at the receiving end of a noiseless binary communications channel (2004, p. 107). Its programs are thought of as code words and the output of the computation as the decoded message. The programs then form what is called a 'prefix-free' set so that successive messages (e.g., procedures) sent across the channel can be distinguished from one another. Despite the different approach of AIT to measuring how much information a message carries, Chaitin acknowledges that AI has precisely the formal properties of SI relative to individual messages (2004, p. 120).

AI may be deemed a competing non-semantic notion of SI by enabling the assignment of complexity values to individual strings and other data types. Like MTC, AIT deals with data encodings of single data structures (typically, strings). So, AI is also information as structured data whose meaningfulness is unaddressed by AIT. Whilst MTC analyses the amount of information of a message relative to

---

[1] The domain of each UTM is self-delimited much like programming languages, since they provide constructs that delimit the start and end of programs. A self-delimiting TM does not "know" in advance how many input symbols suffice to execute the computation (Calude, 2002, pp. 34–35).

a set of messages, AIT analyses the complexity of a string as a single message (Adriaans, 2008, p. 149). The relative frequency of the message, which is the focus of MTC, has no special import in the case of AIT. The length of the shortest program producing this message is minimal within an additive constant that encapsulates the size of the particular UTM[2] representing the amount of information in that message (Calude et al., 2011, p. 5671).

Yet, some have argued that AI may be used to capture the exact amount of "meaning" in a string. For instance, the notion of facticity has been introduced as a measure of how "interesting" or "useful" some string x is. The facticity of x is informally defined as the amount of self-descriptive information that x contains (Adriaans, 2012b, p. 3). This definition is arguably compatible with the intuition that genuinely random strings are not "meaningful" and accordingly their facticity equals 0. On the other hand, a compressible string x, whose AI complexity is smaller than the length of x, has a positive faticity. On this line of argument, despite the common classification of AI as a non-semantic notion of information, AIT might make it possible to also quantify the "usefulness" of information.

In a manner similar to SI, AI too is related to thermodynamic entropy. AI content may also be termed algorithmic entropy (Chaitin, 1990, p. 117). For any computable probability distribution, the expected value of AI equals SI up to a fixed constant term that only depends on that distribution (Li & Vitányi, 2008, pp. 602–608; Teixeira et al., 2011). Also, it has been recently shown that not only is algorithmic entropy analogous to thermodynamic entropy as a measure of disorder defined in statistical mechanics, but it is a special case of the latter (Baez & Stay, 2010), which is typically associated with SI. Algorithmic entropy may simply be defined as the information gained upon learning a number, on the assumption that this number is the output of some randomly chosen program.

6.2.3 Semantic Conceptions of Information

The semantic conception of information is closer to the notion of *meaning* that may be conveyed by a message. It involves a reference to other objects or events in the environment, within a system, in the future, in the past or possibly to things that have never existed and never will (Chaitin, 1990, p. 117). This type of information is typically referred to as *semantic content*. Yet, unlike in the case of quantitative conceptions of information, whether or not semantic content qualifies as semantic information proper remains controversial.

A well-known theory of information as semantic content was developed by Yehoshua Bar-Hillel and Rudolf Carnap. "In distinction to [MTC]", this theory treats "the concept of information carried by a sentence within a given language system [...] as synonymous with the content of this sentence, normalized in a certain way" (Bar-Hillel & Carnap, 1952, p. 1). Bar-Hillel and Carnap's theory of

---

[2] UTMs differ in implementation resulting in the informational content of a string being relative to the particular UTM used to calculate its AI complexity, K. Cristian Calude shows that for every two UTMs u_1 and u_2 $\forall x \exists c$: $(x \in S, c \in \mathbb{N}^+)$ $|Ku_1(x) - Ku_2(x)| \leq c$ where x is the input to the UTM (and S is the set of all strings) (2002, p. 38).

information (BCTI) explains the *meaningfulness* of information relative to some logical probability space. According to BCTI, information is assigned to messages about events and the selected information measure depends on the logical probability of events the message is about. A message is a *proposition* about a situation, property of an object or that a definite event has occurred (Burgin, 2010, p. 321). Logical probability is defined in this context as a function of the set of possible worlds a proposition rules out. The principle underlying BCTI is that the less likely a proposition is, the more informative it is. It is akin to Popper's principle of empirical information as degrees of falsifiability. "[T]he amount of empirical information conveyed by a [set of propositions...] increases with its degree of falsifiability" (Popper, 2002, p. 96).

However, since BCTI offers a *proposition-centred* conception of semantic information, we examine an alternative *data-centred* conception of semantic information. The standard description of semantic information (SDI), characterises information in terms of data + meaning (Floridi, 2005, pp. 351–359) in accordance with the common definitions offered in Section 6.2. According to SDI, an object *O* is an instance of semantic information, understood as semantic content, *iff O* satisfies the following three requirements (Floridi, 2010, pp. 22–23).

1. *O* consists of *n* data (n ∈ $\mathbb{N}^+$). Semantic content cannot be dataless. For instance, a database search query that returns a negative answer, such as 'no entries found', yields some other positive data through either explicit negative data or some metadata about the search query.
2. The data are well formed. Semantic content depends on the occurrence of both syntactically well-formed patterns of data and physically implement*able* differences.
3. The data are meaningful[3]. The question here is *whether* data conveying semantic content can be rightly described as being meaningful independently of the recipient.

Of course, to make sense of SDI, one should first understand what 'data' means. A simple definition of 'data' as physical symbols that are used to represent information is circular, and, hence, does not advance the present analysis of information. Luciano Floridi defines a piece of data (i.e., a datum) as *lack of uniformity* in the world (2010, pp. 22–23). Examples of such lack of uniformity are the presence of some noise, a light in the dark, a black dot on a white page, the difference between the presence and the absence of some signal and a binary 1 as opposed to a binary 0. Data that are processed by some "information agent" can be viewed as "constraining affordances" (Floridi, 2011, p. 87). They are *affordances* for the processing agent to the extent that she can capitalise on them. But they are also *constraints* depending on the interaction with, and the nature of the agent. More specifically, a datum is the lack of uniformity between at least two uninterpreted variables that are distinct from one another in a domain that is left open to further interpretation (Floridi 2011, p. 85). At some level of description,

---

[3] This principle tacitly assumes the existence (even in the past) of some agent (possibly, with a system of values) relative to whom the data are (or were) meaningful.

this distinction may be the lack of uniformity between at least two signals, such as a dot and a dash (in the domain of Morse code) or a variable electrical signal (the domain being a phone conversation). But, importantly, at another level of description, those signals make possible the coding of some distinguishable symbols, say, α and β in the Greek alphabet.

Whether or not semantic content, as defined by SDI, is sufficient as an analysis of declarative semantic information is controversial. The key question is whether semantic content qualifies as declarative semantic information even when it does not represent or convey a truth. If the answer is affirmative, then misinformation and disinformation also count as genuine types of declarative semantic information (Dretske, 1981, pp. 44–45; Floridi, 2005, pp. 359–360). Some philosophers have argued that semantic content need not be truthful to qualify as declarative semantic information (Bar-Hillel & Carnap, 1952; Fetzer, 2004; Scarantino & Piccinini, 2010).

Consider the consequences BCTI's analysis of information. According to BCTI, the less likely a proposition is, the more informative it is. It follows that tautologies convey 0 information for they have an "absolute logical probability" (Bar-Hillel & Carnap, 1952, p. 2) of 1. On the other hand, logical contradictions, whose probability is 0, are *maximally* informative, for the very same reason, despite being false (Bar-Hillel & Carnap, 1952, pp. 7–8). According to Bar-Hillel and Carnap, this consequence is not problematic, since on their view, declarative semantic information does not imply truth. Floridi argues that these consequences are paradoxical and serve as a reason for insisting on the veridicality of declarative semantic information. According to BCTI, by making a proposition p less and less likely, the logical probability of p becomes zero when p is logically impossible. But, counterintuitively, at that point p becomes maximally informative (Floridi, 2010, p. 58), as it rules out the set of *all* possible worlds. To avoid such paradoxical results, Floridi endorses the Veridicality Thesis, according to which truthful information is determined by an information-to-world direction of fit. A signal s carries the information that p iff s has the semantic content that p relative to a factual event in the world. So, to be compatible with the Veridicality thesis, by Floridi's lights, SDI has to be extended to also include the following fourth requirement.

The Veridicality Requirement: The data are true (Floridi, 2005, pp. 365–366).

With this background in mind, it is easier to understand the difference between the two semantic conceptions of information examined as candidates for an IP account of computation. The first one is *declarative* semantic information and the other is *prescriptive* semantic information. The former describes objects, events or states of affairs in the world and as such it has an alethic nature; it can be evaluated as either true or false. According to Floridi, only when it is true does it qualify as *genuine* descriptive information, which he calls *factual semantic information*, and avoids the paradoxical consequences of BCTI (2011, p. 239). The latter, on the other hand, concerns relationships that specify a course of action.

One type of semantic information that is examined here, as the basis of an IP account of computation, then is *instructional* information. Instructional information is prescriptive in that its processing function is aimed at making something happen. It is not about some fact or state of affairs, so it cannot be correctly qualified as true or false (Floridi, 2010, pp. 35–36). An instruction manual for an oven contains instructional information for producing the expected result of either cooked or warm food. Consider the following piece of information: "Turn the temperature knob to '200' to set the oven to 200° Celsius". The instruction "Turn the temperature knob to '200'" is neither true nor false irrespective of whether following it yields the desired state of affairs (i.e., a temperature of 200° Celsius in the oven).

Unlike the case of factual information, whilst the three requirements of SDI are necessary for instructional information, the Veridicality requirement is not. Instead, it is the correctness and lack of ambiguity of the semantic content that contribute to yielding the right state of affairs when instructional information is processed. This is reflected by the following supplementary requirement.

The Prescriptive Requirement: The *satisfaction* of semantic content c has to yield a definitive action in a given context for c to qualify as instructional information.

Thus characterised, it is easy to see that instructional information provides directions for action. In philosophical terms, instructional information may be described as yielding the knowledge-*how*, under the right conditions, for accomplishing some task, rather than the knowledge-*that* of the task being accomplished. This characterisation accords with imperatives, which are at the heart of practical and moral reasoning. One reductionist attempt at developing imperative logic translates *imperatives* into *declaratives*. Yet, no such reduction seems to fully encapsulate the meaning of imperatives (Green, 1998). The core feature of imperatives is that an imperative is satisfied *only if* the action it enjoins is brought about (Hamblin, 1987, p. 45). This feature is reflected in the Prescriptive requirement. Importantly, for our purposes here, instructional information can be conveyed either conditionally ("If it rains, shut the window!") or unconditionally ("Shut the window!").

The other type of semantic information that is examined as the basis of an IP account of computation is *factual* information. One obvious question is why the BCTI conception of semantic information is not examined instead of factual information. For the Veridicality requirement is clearly problematic. This is true and yet, the BCTI conception is *proposition*-centred rather than *data*-centred. It is the latter type that seems to be a more natural candidate for an IP account of computation and it accords with the Data – Information – Knowledge triad. Still, the particular problems that an IP account of computation based on factual information faces due to the Veridicality requirement are examined below.

6.3 Features of the Resulting IP Accounts Based on the Different Conceptions of Information

6.3.1 A Possible Objection

Before examining features of the resulting IP accounts, let us pause briefly to consider a possible objection that might be raised at this point. According to this objection, the classification of SI and AI as non-semantic conceptions of information and factual information and instructional information as semantic conceptions is misguided. For one thing, according to BCTI, semantic information does not encapsulate truth, and the underlying definition of logical probability was the starting point for Solomonoff's notion of a universal distribution. His work on AIT was motivated by philosophical questions concerning the nature of probability and the induction problem, associated with BCTI (Adriaans & van Benthem, 2008, p. 8). The intimate relation between BCTI and AIT suggests that classifying AI as a non-semantic notion of information is problematic (cf. the notion of facticity briefly mentioned in Section 6.2.2).

Additionally, factual information, so the objection continues, does not have a particular import for the underlying concepts that are studied in computer science. The concept of information as a subject of rigorous philosophical investigation is accessible to us via its various mathematical formulations, such as MTC, AIT and Gibbs entropy[4] (Adriaans, 2010). Whilst the Data – Information – Knowledge triad (more on this is said in Section 6.4) may have an intuitive appeal in the context of the present discussion, it is supposedly shaky as a foundation for a philosophical analysis of information. For the age-old debates about the necessity of truth for both information and knowledge become unavoidable. Further, fundamental problems, such as what the relation is between digital computation and the growth and discarding of information, are arguably inexpressible in the vocabulary of factual information.

For a complete reply to these challenges, the reader is encouraged to read the entire chapter to glean applicable results pertaining to these various claims. Nevertheless, let us make a few remarks here. To begin with, before concluding that factual information is indeed inadequate as a candidate for an IP account of concrete computation, some groundwork is needed. We cannot simply dismiss it, since some CTM take computation to be the processing of *semantic* information. For example, according to both the PSS and the FSM accounts, digital computing systems fundamentally process semantic information in such a manner that this processing preserves the truth of the processed information, thereby ensuring that it continues to correspond to some external state of affairs (cf. the fifth requirement of the PSS account and the first requirement of the FSM account discussed in Chapter 5).

---

[4] Gibbs entropy measures the amount of disorder in a closed system of particles in terms of probability distribution of the particles allowing to make inferences based on incomplete information (Bais & Farmer, 2008, sec. 3.5).

Besides, the relations that obtain among semantic information, SI and AI can certainly not be denied. Despite there being no explicit agency involved in the MTC analysis of SI, unlike the case of data-centred semantic information, these two conceptions are not mutually exclusive. For one thing, coding systems relate to the efficiency of *natural* language and signal probability relates to the reliability of sources that is important in logic. Additionally, question-answer scenarios are often used in SI-based analyses to pin down some "truth" by using the minimal number of questions. Similarly, as regards AI, if an agent has background knowledge, in the form of optimal descriptions of some set of objects, then identifying such an object via, say, a picture amounts to finding a shortest algorithmic description of the picture conditional on that background *knowledge* (Adriaans & van Benthem, 2008, pp. 11–13).

Whilst mathematical conceptions of information are certainly invaluable when quantifying the information processed, their applicability as candidates for an adequate IP account is debatable as is argued below. Having a measurable *amount* of information as calculated by either MTC or AIT does not, in itself, allow a string to express a question or an instruction (Sloman, 2011, pp. 398–399). Yet, instructions and questions are precisely the ingredients that seem to be required for an adequate explanation of computation (*online* computation, in particular). As such, at least prima facie, *instructional* information is a better candidate for an IP account than SI or AI. As we shall see presently, an IP account based on *instructional* information has the required backbone for explaining *what* computation is.

As well, subscribing to such an IP account does not preclude the use of MTC or AIT as tools for tackling some of the fundamental problems in computer science. Endorsing the view that digital computation *is* the processing of instructional information still permits the much-needed analysis of, say, the intimate relation between computation and the growth and discarding of information. This aspect of computation as information processing is discussed as one of the key requirements for a physical system to digitally compute. Instructional information, so it is argued below, offers the right level of abstraction for describing digital computation, without thereby hindering the mathematical explanatory power that comes with either MTC or AIT. With that in mind, let us now turn to examine features of the resulting IP accounts of computation.

6.3.2 Features of the Resulting IP Accounts Based on Non-semantic Information

An SI-based IP account is, at best, limited in its ability to explain discrete deterministic computation in physical systems. Such an account clearly has some merit for explaining concrete computation. MTC emphasises the role of symbol structures as designating a particular state of affairs out of some larger set of possible states (i.e., selective information). SI is fundamentally a non-actualist conception of information, for it considers the actual world as simply one of many other possible worlds.[5] For the actual message transmitted is just one of many other possible messages. To a great degree, conventional digital computers are non-actualists, in the sense that they are designed to resist minor perturbations (e.g., noise) and respond to a broad class of inputs. The non-actualist character of digital computing systems suggests that there exists some compatibility between them and SI.

Indeed, the selective characteristic of SI is compatible with a particular control structure enabling the remarkable flexibility of programmable computing systems. Selective information is closely connected with the way in which digital computers are capable of performing a conditional branch. This operation detects which of several different states obtains (e.g., what input was received by the computer program or which symbol was last scanned by the TM) and then sends the computation along different paths accordingly. The use of selective information by conditional branch processes lies at the heart of everything complex that digital computers do (Ralston, 2000, p. 856).

Furthermore, hardware malfunctions in deterministic computing systems can be described as noise in discrete communication channels. MTC deals with those aspects of communication where a transmitted signal is perturbed by noise during transmission. The received signal is then analysed as a function of the transmitted signal and a probabilistic noise variable (Shannon, 1948). When considering a miscomputation, which is the result of a hardware malfunction, as a malformed signal, an analysis of it in terms of SI processing can be useful. The computer's memory registers, for instance, are designed to handle such noise by including error correction mechanisms, such as using parity bits and information redundancy. An even more obvious example is a discrete connectionist network in which one of the neurones has lost the ability to transmit and/or receive signals to/from some neighbouring units. Such a scenario can be analysed in terms of noise and the channel's capacity to send and receive messages.

Yet, SI is a probabilistic concept whereas digital computation may be either deterministic or not deterministic (e.g., probabilistic computation or pseudo-random computation). The description of a physical system in terms of SI is only adequate if it is memoryless, that is, if the system's transition to state S_{j+1} is unrelated to its last state S_j. But this is typically not the case for most conventional digital computing systems, which are deterministic, for their behaviour, in principle, is repeatable and systematic, as discussed in Chapter 3.

Whilst the state-transitions of Shannon's communication model are probabilistic, the transition probabilities of a *deterministic* TM are all set to 1. For every possible valid input and a given initial state, there is only one possible state into which the TM transitions. Every future state transition can be accurately predicted by simulating the program being executed. So, in the case of idealised TMs, there is no element of uncertainty or surprise, on which SI is fundamentally based. Nevertheless, in the case of conventional digital computers, which are susceptible to noise (at both the software and the hardware levels), an SI-based IP

---

[5] This is consistent with the possibilist's thesis, in the metaphysics of modality, that the set of all *actual* things is only a subset *all* of the things that *are* possible.

account may be useful in describing such potential adverse effects on otherwise deterministic computations.

Let us pause for a moment to preempt a possible objection to examining the features of an AI-based IP account. An AI-based IP account of computation, according to this objection, inevitably yields a circular definition of computation. For AI is defined in terms of computational programs and the very computing systems that we seek to explain. This is a fair criticism of an analysis of AI as a candidate for an IP account of computation. In a sense, it weakens the claim above that mathematical formulations of information are those that enable a rigorous philosophical investigation needed for understanding computation as information processing. Nevertheless, despite the potential circularity inherent in an AI-based IP account, we may at least be justified in wondering what would be the merits of such an account. We can now turn to the task at hand.

In a manner similar to SI, AI is a non-actualist conception of information. In accordance with the Universality Theorem[6], any program in some computer language can be converted into a "program" running on a UTM. There is some algorithmic procedure for passing among the possible enumerated TMs that compute a function f. To that end, one can pretend that she has all these enumerated TMs in front of her. If, for instance, one needs to compute 20 steps in the computation of the 5th machine on input string "slam dunk", then the 5th machine is selected, "slam dunk" is inscribed on its tape and that TM is run for 20 steps (Downey & Hirschfeldt, 2010, p. 9).

In order to assign AI complexity to the configuration of some computing system (both the machine and the self-delimiting program it simulates), the system is "frozen" at some point in time. The snapshot of the *actual* computation that takes place, rather than all possible counterfactual computations, is assigned an AI complexity value. Nevertheless, since all the enumerated TMs for computing f are available *in principle*, the AI complexity analysis is not strictly limited just to the actual computation that is executed.

Whilst an SI-based IP account can only tell us what the lower bounds for solving a given computational problem are, an AI-based IP account can distinguish among different optimal programs for solving a specific problem. MTC provides mathematical measures to calculate the *lower* bounds of information flow along a channel. It can tell us that a solution to a given problem cannot be computed in less than n bits of information. But an SI-based analysis cannot distinguish between two equally small circuits or different optimal programs for solving the same problem (and there are infinitely many such programs). On the other hand, in the context of AIT, the set of all such possible optimal programs is enumerable, in principle (Calude, 2009, p. 82). The full description of each one of these enumerated programs can be provided using the AIT framework. But traditional AIT can only *approximate* the complexity of a string relative to a particular UTM and this prevents us from actually having a full

---

[6] The Universality Theorem states that there exists a self-delimiting UTM U, such that for every self-delimiting TM T, a constant c can be computed (depending only on U and T), satisfying the following property. If $T(x)$ halts, then $U(x') = T(x)$, for some string x' whose length is no longer than the length of the string x plus c (Calude, 2009, p. 81).

description of all the optimal programs producing that string (Calude et al., 2011, pp. 5668–5669).

A variation on traditional AIT, which is not based on UTMs, allows us to compute the complexity of strings (or the exact length of these optimal programs), though this comes at a cost. Finite State Transducer AIT (FSTAIT) relies on finite transducers for which the universality theorem is false. (To reiterate, a transducer is an FSA with outputs). A transducer can be described as the 6-tuple $(Q, \Sigma, \Gamma, q_0, Q_F, E)$, where Q is the finite set of states, Σ and Γ are the finite sets of input and output symbols respectively, $q_0 \in Q$ is the initial state, $Q_F \subseteq Q$ is the set of accepting states and E is the finite set of transitions (Calude et al., 2011, p. 5669). Since there is no universal transducer, FSTAIT cannot achieve universality.[7] At the same time, the finite state complexity of a string explicitly includes the size of the particular transducer running the program and this complexity becomes computable. Traditional AI complexity can only be approximated. For our purposes, this means that different optimal programs are not only enumerable but also *distinguishable* from one another.

An AI-based analysis is closely coupled to the implementing machine used to run the program in question. The machine's size is included as part of the encoded length of the computed string (Calude et al., 2011). When AIT examines which problems can be solved, either a *particular* self-delimiting UTM or a specific combination of finite state transducers (in the case of FSTAIT) is considered.

6.3.3 Features of the Resulting IP Accounts Based on Semantic Information

We first examine features of the resulting IP account that is based on *factual* information. On such an account, any deductive-like processing of factual information is both *representational* and, at least in principle, *truth preserving*. Symbolic representations, or, say, sub-symbolic representations in connectionist networks, that are processed in the course of a computation, supposedly, carry *true* information about some, possibly external, state of affairs. Since factual information represents objects, events or states of affairs in the world, the resulting IP account is committed to *extrinsic* representations.

Unlike SI and AI, factual information is an *actualist* conception through and through. The preservation of the isomorphic mapping between a representation and the represented state of affairs requires that their processing be truth preserving. Factual information, by Floridi's lights, refers to situations that either

---

[7] There is no finite generalised transducer that can simulate a transducer running some program. Yet, Calude et al. (2011, p. 5672) prove that the Invariance Theorem (informally, stating that a UTM provides an optimal means of description up to an additive constant) also holds true for finite state complexity. Finite state complexity of a finite string x is defined in terms of a finite transducer T and a finite string s such that on input s T outputs x. It is defined relative to the number of states of the transducers used for minimal encodings of arbitrary strings.

have been actualised or are actualising in the world. A factual assertion commits the asserter not to its truth in some *possible* world but in the *actual* world. This already suggests that processing of factual information and concrete digital computation are incompatible, for the latter is a non-actualist phenomenon.

This incompatibility raises a problem in regard to concrete computation in human-engineered systems. A digital computing system allows all the *possible* computations, which the system *could have* performed, rather than just the one it *actually* performs. This point was used to resist Searle-like trivialisation of computation as discussed in Chapter 4. An adequate account of computation should not be confined only to the actualised computation. However, an IP account based on factual information can only consistently explain actualised computation, due to the Veridicality requirement and the constraining causal relation to represented object(s) or event(s) in the world.

On the other hand, an IP account based on instructional information seems less problematic. A program executed on a conventional digital computer can be construed as the execution of instructional information (e.g., assign the value '3' to variable var1, do X if Y, otherwise halt; where X stands for an instruction and Y is the condition). An ALU can be described in terms of executing instructional information depending on the number of arithmetic and logical operations it supports. Similarly, the working of some discrete connectionist networks can be explained in terms of instructional information, if they perform two or more operations (e.g., both n-bit disjunction and conjunction operations). Other discrete connectionist networks simply process data that does not rise to instructional information, since instructions are not necessary for their normal operation.

Additionally, instructional information is not limited to specifying only *actual* computations. The informational content of a program is given by its behaviour on all *possible* inputs, subject to some practical restrictions, rather than just on *actual* inputs. Further, data and instructions may be interchangeable in the context of executed programs. UTMs, and similarly general-purpose digital computers, may take the same vehicles (be that strings, numerals or anything else) as either instructions or non-instructional data. The very same string may play the role of an instruction in one run of the program and non-instructional data in another (Scarantino & Piccinini, 2010, p. 326). The same string may even be both an instruction and data in a single run of a program. Instructional information is certainly compatible with this principle. There is nothing in the definition of instructional information that implies that this information only applies to actual occurrences, for it is not evaluated alethically.

6.4 The Key Requirements Implied by the Resulting IP Account

We now examine five key requirements for a physical system to perform *nontrivial* digital computation in terms of information processing in the context of the four conceptions of information discussed above.

- 1st Requirement: having the capacity to send information.

- 2nd Requirement: having the capacity to receive information.

- 3rd Requirement: having the capacity to store and retrieve information.

- 4th Requirement: having the capacity to process information.

- 5th Requirement: having the capacity to actualise control information.

The first requirement is the system having the capacity to send information. This requirement is implied by all four resulting IP accounts. The sender prepares the messages to be sent to the receiver and encodes them for transmission. An important distinction that should be drawn in the context of digital computing systems is between sending information *internally* amongst different components of the same system and *externally* between the system and some external interface (e.g., an output device, a transducer or another computing system). A computing system devoid of any external interfaces may still compute a solution to some specifically predefined problem. An example of sending information internally is the main memory of a conventional computer being the *source* of information (e.g., a stored instruction). The memory *controller* acting as the *sender* is responsible for fetching data from the main memory and transmitting them to the CPU. Similarly, the TM's tape can be regarded as the source of information when its head, acting as the sender, reads a symbol from the tape.

Importantly, there is an upper bound on the velocity of message transmission. The existence of this bound was emphasised by Gandy: "[there is a] finite velocity of propagation of effects and signals" (1980, p. 135) (see Chapter 7 for more details). In contrast to Newtonian mechanics, Einstein's theory of relativity dictates a finite bound in nature on the velocity of transmission of physical signals that is equal to the speed of light (approximately equal to 3×10^8 m/sec) (Okun, 2009, p. 29). In Newtonian mechanics, where gravitation plays a key role in the structure of spacetime, physical signals can propagate faster than the speed of light. Nevertheless, no physical actions have been discovered yet that are actually transmitted faster than light (Bohm, 2006, p. 189).[8] The upper bound on the velocity of signal transmission has two immediate consequences. The first one is that a sender can only transmit information within a bounded region in a limited time interval. The second one in the context of deterministic digital computing systems is that there exists an upper bound on the number of possible computational steps per second, since each step depends on the complete configuration of the system.

The second requirement is the system having the capacity to receive information. This requirement is most obvious in MTC. In the context of computing systems, this requirement needs to be relaxed a little. If the former requirement necessitated a sender to transmit the message, this requirement expects a receiver on the other end to accept it, at least in principle. In some cases,

---

[8] Quantum computing systems introduce some potential complications in this context, particularly, concerning bounded regions. Nevertheless, their treatment exceeds the scope of the present work.

the absence of a receiver on the other end means that the computation is either unexecuted or suspended. For instance, a program thread sending an input/output signal to the OS (acting as the receiver) will enter the suspended mode until its input/output request is acknowledged. But whereas a sender is needed to transmit the information, there are cases where the absence of a receiver does not result in an incomplete computation. This is particularly common in intercomputer communication, but applies to individual computing systems just the same.

One obvious example is certain communication protocols invoked by a computing system. Some communication protocols such as TCP (Transmission Control Protocol), which is at the heart of all HTTP-based Internet transactions, require a "handshake" between the sender and the receiver for the transaction to be successful (e.g., the "Server not found" error message displayed when a particular website is unreachable). But other protocols such as UDP (User Datagram Protocol), which are less reliable but faster, do not require that the receiver acknowledge the receipt of the message sent.

The third requirement is the system having the capacity to store and retrieve information. It is not implied by all four resulting IP accounts. Notably, SI is based on a memoryless source of information and a memoryless channel of communication. A message sent at time t_2 is statistically independent of the previous message sent at time t_1. AI, on the other hand, at least implicitly presupposes memory, for AIT relies on computing systems that use memory (e.g., a UTM). Semantic content is wedded to memory due to the third requirement of SDI. The meaningfulness of data requires memory, for it provides the continuity between the past, present and future states of the sender(s) and receiver(s). We conclude, therefore, that both instructional information and factual information presuppose memory as well.

To make the discussion of memory storage and retrieval more precise, we note that some computing systems are stateful, whilst others are not. Stateful computing systems are those whose operation at time t_2 depends on their state at time t_1. A TM is a paradigmatic example of an abstract computing system whose next operation depends on its current state. There are also obvious examples of stateful physical computing systems, such as flip-flops and shift-registers. A flip-flop is a relatively simple circuit that has two stable states, which can change by applying signals to at least one of its control inputs. A shift register, which is a set of flip-flops, is used to store binary data and can be viewed as embodying a bit array. The result of its bit-shift operation at time t_2 depends on the bits that were stored in the register at time t_1. Basic logic gates, on the other hand, are not stateful. For example, an AND gate performing logical conjunction at time t_2 does not depend on its state at time t_1. Each logical conjunction operation is performed independently of the last operation.

This requirement also does not apply uniformly to all discrete connectionist networks.[9] Recurrent networks change their persistent state(s) in the

---

[9] Arguably, the training of, say, a feedforward network produces some "memory": the network remembers what it has been trained to do. However, once the network is fully trained, it no longer retains any new data it receives. We only consider fully trained connectionist networks here.

course of their operation and thereby qualify as memory-based systems. The memory capacity of recurrent networks is enabled by feedback loops by which a particular neurone is connected to itself either as an input neurone or through a series of edges. The output of such a neurone at time t_1 can affect its activation state at time t_2. Feedforward networks, on the other hand, do not have such feedback loops. Their operation proceeds in a layerwise manner starting from the input layer through the hidden layers, if there are any, to the output layer. The signals received at the input layer propagate all the way to the output layer in a single pass through the network without any neurone's activation state being stored.

The fourth requirement is the system having the capacity to process information. It is the most problematic requirement and it becomes even more stringent if information is construed as *factual* information. To reprise, the processing of information is characterised as encompassing the encoding, decoding, production, modification and removal of information.[10] The production of new information, such as a new database table containing salaries of employees[11], may be the derivation of new propositions from existing ones. The modification of existing information, such as giving some employees a pay rise, is the manipulation of some information I_1 such that $I_1{\neq}I_2$ (where I_2 is I_1 that is modified at T_{n+1}). The removal of information, such as deleting matching records of employees, who left the company, is a selective removal of information from the system that need not result in the deletion of the entire system.

The processing of information then is broader than merely encoding and decoding information. Strictly, it should not be equated with the *transformation* of information either. While transformation may be the modification or removal of some information, it does not imply the production of *new* information. The transformation of information implies some prior form of information changing to another form.

Arguably, deterministic computational processes have a limited capacity to produce new information, and over the course of time they discard information (Adriaans & van Emde Boas, 2011; Calude, 2009, pp. 84–85). The discarding of information about the history of the computation and the associated energy dissipation was originally conveyed by Landauer's principle of logical irreversibility (Landauer, 1961). According to this principle, the erasure of information in the course of computation requires energy dissipation, which has an absolute minimum that is independent of the rate of the computational process. This unavoidable minimum is due to the "fact that the computer performs

---

[10] Information processing may be construed in a variety of ways depending on the particular context of enquiry, including (but not limited to) the manipulation, acquisition, parsing, derivation, storing, comparison and analysis of information. However, it seems that these operations depend crucially on at least one of the aforementioned operations.

[11] Yet, it remains to be seen whether this new information stored in the database of a digital computing system is merely a *copy* of the new information that was produced externally (e.g., by the human resources manager).

[logically] irreversible operations" (Landauer, 1961, p. 190).[12] Landauer's principle is a particular formulation of the second law of thermodynamics in the context of information theory. It "expresses the fact that erasure of data [and, therefore, information] in a system necessarily involves producing heat, and thereby increasing the [thermodynamic] entropy" (Bais & Farmer, 2008, p. 633). Both SI and AI can be used to convey this increase in thermodynamic entropy.

What is the difference between logically reversible and irreversible operations? A logical operation takes a finite number of distinct input states and maps them to a finite number of output states. A logical operation for which an output state has more than one possible corresponding input state is irreversible. A logical operation for which each output state has precisely one possible corresponding input state is reversible. Given any output state it is always possible to determine its corresponding input state.

Most conventional computers perform irreversible computations. Consider first the calculation of $x + y = z$ (for $x, y, z \in \mathbb{N}^+$). In the process of transforming the input x, y to the output z, some information is lost. Since, in general, there are many possible pairs of addends adding up to the output z (for $z>3$), an algorithm that computes the function $f(x, y) = x + y$ discards the information about the precise history of the computation. Consider next, the computation of an irreversible two-input, one-output AND gate. When the gate produces the equivalent of a logical '0' (represented by the appropriate voltage level), it discards the information regarding the original values of its two input lines. Since all three input combinations (0, 1), (1, 0) and (0, 0) yield the same output, the history of the logical AND operation is discarded, once the logical '0' is produced as output. On the other hand, the negation operation, performed by a NOT gate, is fully reversible.

Let us consider some examples at the two extremes of the spectrum of information production and removal in computer programs. In pure information discarding computations, such as the summation of a set of numbers or the deletion of a string, the program typically reaches an accepting state by means of systematically reducing the input. In pure counting computations, such as in "for int i=0 to n write('1')", the AI complexity of the input does not grow as a function of the size of the counter, since n is not explicitly defined (we may assume it takes values between 0 and, say, 10000) and the program length does not change.[13] Lastly, in pure search computational processes where the input is not reduced, the AI complexity of the input remains constant[14] for the information is kept unchanged throughout the whole process (Adriaans & van Emde Boas, 2011, p. 15).

---

[12] It has since been argued that logical irreversibility can be avoided in discrete computation in general (Bennett, 1973). The implication of replacing logical irreversible operations with reversible ones is an increase in storage memory for keeping the history of the computation.

[13] Of course, this is not the case when one considers the difference between "for int i=0 to 10 write('1')", say, and "for int i=0 to 10000 write('1')". Here, as the input size grows, the size of the counter grows logarithmically. The AI complexity of the second FOR statement is bigger than the size of the first.

[14] If one insists that the location of the searched object in the search space constitutes new information, then the amount of this new information is at most upper-bounded by a constant (Calude, 2009, p. 85).

Be that as it may, the production of new information and modifying information requires the ability to distinguish between different contents in the context of computational processes. MTC tells us about the probabilities associated with symbols from a given language. The processing of SI may be viewed as the modification of the conditional entropies among the states of signals encoding the possible messages. But MTC is indifferent to the *content* of the messages. For instance, the following strings S_1 and S_2 have the same length (including that of their symbol constituents). S_1="All cars have four wheels"; S_2= "All cats have four ankles". Let us suppose that S_1 and S_2 are equiprobable (so according to MTC, they are equally informative). Let S_3 be "Bumblebee is a car". By using Universal Instantiation and Modus Ponens (taking these as being represented in first-order predicate logic), one can infer some new justified information.[15]

What does this example show us in the context of processing SI? One can infer, for example, "Bumblebee has four wheels" (=S_4) from S_1 and S_3. The overall informative content of any two different and independent strings combined is typically greater than that of each one of these strings individually.[16] This new information tells us something else about Bumblebee: it has four wheels. S_2 and S_3, however, do not yield new justified information using Universal Instantiation and Modus Ponens, as in the case of S_4. One cannot validly infer any new singular statement about Bumblebee from the universal statement S_2. In order to apply rules of logic as a means of producing new (true) information, the symbolic constituents of strings must be distinguishable. But according to MTC, we may encode and transmit S_2, rather than S_1, and S_3 to the recipient (as S_1 and S_2 are equiprobable) that learns nothing new from S_2 and S_3 in this case. The processing of SI is indifferent to the recipient learning something new by receiving S_1 and S_3 (and using rules of logic) as opposed to receiving S_2 and S_3 given a uniform probability distribution of these messages.

When information is construed as *factual* information its processing requirement becomes even more stringent. The syntactical manipulation of messages has to be done in a manner that preserves their semantics and by

---

[15] There is an ongoing debate regarding information in deductive inferences. Some, including John S. Mill and the logical positivists, have argued that logical truths are tautologies, and so deductive reasoning does not produce any *new* information. On this view, all valid deductive arguments simply beg the question. Others, notably, Jaakko Hintikka (1984), have argued that deductive reasoning can indeed produce new nontrivial information.

[16] Generally, the amount of information in any two strings Si and Sj is no less than the sum of the information of Si and Sj, if the content of Si and the content of Sj are in some sense independent or at least one does not contain the other. Still, there are clearly cases where the informational content of their sum ($Si + Sj$) is less than the sum of their informational content individually. For a more detailed discussion of the "additivity" principle see (Bar-Hillel & Carnap, 1952, pp. 12–13). Arguably, Universal Instantiation and Modus Ponens, for instance, as a means of inferring S_4 from S_1 and S_3 also carry some positive information, since without the recipient knowing how to use them, she cannot infer S_4.

implication the truth of the data processed. At the very least, new justified information has to be consistent with prior existing factual information. If conjunction, for instance, is applied to produce new justified information, then the conjuncts C_1 and C_2 must be neither contradictories nor contraries. Otherwise, their conjunction $C_1 \wedge C_2$ would be false and, thus, non-factual.

A database-driven computing system that progressively produces new information is a useful example. Consider a system C_S whose database is initially populated with some basic propositions and is designed to progressively increase the database's overall information. If the information processed by C_S is taken to be factual and, hence, true, then the resulting new information has to be true as well. C_S can progressively produce more information by means of logical inferences. For instance, if some propositions P and Q were entered initially, then C_S could produce the new proposition $(P \wedge Q)$ and add it as a new entry in the database. Intuitively, the more propositions there are in the database, the more informative it becomes (again, provided that deductive inference can produce new, even if limited, information). But this requires that C_S be capable of determining which propositions are true and which are false. Otherwise, inconsistencies may eventually creep into the database, thereby decreasing its overall informative content (though on BCTI, such inconsistency yields maximal information).

The fifth requirement is the system having the capacity to actualise control information. It is only strictly implied by instructional information and presupposed by AI. Whilst the four preceding requirements apply to many information-processing systems, they are insufficient for distinguishing trivial computing systems from nontrivial computing systems. Basic logic gates, half adders and full adders send, receive and transform[17] information, but they only perform *trivial* digital computation. This already reveals the adequacy of the four different conceptions of information as the basis for an IP account of computation.

Trivial computing systems lack the capacity to actualise control information. Logic gates can be wired together to form computing systems, whose computations can be logically analysed into the operations performed by their components. This is, after all, how computers as finite state machines are built. Nevertheless, not every collection of entities, even provided that they may be described in isolation as logic gates, can be connected together to form a digital computer proper. The inputs and outputs of logic gates are typically of the same type, so that outputs from one gate can be transmitted as inputs to other gates. The components have to be appropriately organised and synchronised (Piccinini, 2007, pp. 522–523), but also include some controller unit(s) to qualify as a digital computer proper.

Control information is characterised as the capacity to control the acquisition and utilisation of information to produce a definitive action.[18] Accordingly, control information is distinguished functionally from the process of *exercising*

---

[17] The reader will have noticed the deliberate use of 'transform' here, rather than 'process'. For information processing, but not its transformation, also implies the possible production of new information.

[18] This characterisation is an adaptation of Peter Corning's teleonomic definition of control information in cybernetic and biological systems (2001, p. 1277).

control (Corning, 2001, p. 1276) and it only exists *in potentia* until the system actually uses it. Control information is always relational and context dependent. It also has no fixed structure or value – a single binary bit may be sufficient for producing a definitive action by choosing between two possible courses of action, such as either *continue* or *stop*.

Control information is content-based and this affects its relation to the four conceptions of information. It is excluded from SI, since MTC ignores the specific content of the information carried by messages. In being structured data devoid of meaningfulness, SI lacks the properties required for discriminating control information given a fixed set of messages. In grounding informational analysis on UTMs (or any other control mechanism, such as an FST), AIT implicitly assumes control information. Still, AI, in linking information with data encoding in a single data structure, lacks the properties required for discriminating control information. It is only instructional information that is fully compatible with control information. Strictly, the former and the latter are arguably equivalent. Factual information, on the other hand, by itself has no similar implications for action (Sloman, 2011, pp. 418–419). It seems then that it is only instructional information that implies the fifth requirement, whereas AI simply presupposes it.

6.5 Problems for the Resulting IP Accounts

6.5.1 Problems for IP Accounts Based on SI or AI

To start with, an IP account of concrete computation that is underpinned by SI has a limited explanatory power concerning deterministic processes. SI only makes sense in the context of a set of potential messages that are communicated between a sender and a receiver and a probability distribution over this set (Adriaans, 2008, pp. 146–147). There is no room for a probabilistic selection of messages in describing deterministic processes, for its probability is 1, barring potential adverse effects of noise as discussed above. Deterministic computation is describable as a *specific set* of messages, which are selected, encoded and transmitted in a certain ordered sequence of steps, without any probabilities associated with these messages. The processing of SI does not entail deterministic digital computation.[19]

AI is an improvement on SI as a basis for an IP account of computation and in some regards does better in terms of explaining deterministic computation. AIT analyses the complexity of a string relative to a particular UTM as a single message and hides the probabilistic message selection process of MTC. AIT (and more specifically FSTAIT) can describe the behaviour of optimal programs. FSTAIT can describe the behaviour of non-optimal programs too, if the size of the

---

[19] In a similar vein, it is argued in Piccinini and Scarantino (2011, p. 33) that the processing of SI does not entail digital computation.

program in bits is specified[20]. The resulting non-optimal programs then become enumerable and distinguishable from one another as in the representative case of optimal programs (Calude, personal communication).

As suggested above, characterising computation as the processing of either SI or AI is problematic. The focus of SI is not on the content of individual messages, but that content *is* precisely what gets *manipulated*. The processing of SI can be the modification of the states of signals encoding the possible messages, the elimination of possibilities, that is, a reduction in uncertainty, represented by a signal or the introduction of redundancy to offset the impact of noise and equivocation. Still, sending the same message twice as a means of introducing redundancy does not yield more information than that in each. It may be argued, though, that increasing the reliability of the message transmission process instills some confidence in the receiver. Even if that were the case, the informational content would not increase further by sending each message, say, three times instead of just twice. Similarly, the elimination of redundancy does not reduce the underlying informational content.

Noise on the channel is the source of modification and removal of information and uncertainty is the source of new information. The modification of SI by using error correction methods is a means of offsetting noise. But even then, the underlying informational content of the messages remains largely unmodified. Reversible deterministic computation that processes SI can only *preserve* SI. Noise that causes the removal of information is typically physical and rarely ever deliberate. We constantly try to find new ways to minimise the adverse impact of noise on communication (e.g., by using parity check bits or Hamming codes). By contrast, the deletion of information in digital computing systems can be completely deliberate, say, to free up memory resources. Irreversible deterministic computation that processes SI can discard SI when a many-input-single-output computational path is followed. The presence of noise in SI-processing computation can also lead, of course, to the discarding of SI. Lastly, new SI is produced only relative to the uncertainty associated with that information. If the entropy of a message in a particular context is 0, then sending this message will not amount to producing new information. Since deterministic computation does not increase uncertainty, it cannot be said to produce new SI.

What does the processing of AI amount to? Producing new information amounts to the system producing an output string S_{OUTPUT} that *encodes* more information than the input string S_{INPUT}. The production of new information amounts to the AI complexity of S_{OUTPUT} being greater than that of S_{INPUT}. Conversely, the deletion of information amounts to the system starting with S_{INPUT} and producing S_{OUTPUT} with *less* information than S_{INPUT}. If data are construed as lack of uniformity, then a complete deletion of *all* data can only be achieved by the elimination of *all* differences, thereby restoring uniformity. A system that

---

[20] Dealing with optimal programs is a feature of at least conventional AIT. But this by no means has any special bearing on AIT being an adequate candidate for an IP account of digital computation. Rather, the point is that AIT, unlike MTC, can adequately describe the behaviour of different programs.

produces S_{OUTPUT} with *less* information than S_{INPUT} means that only some unwanted information is discarded.

Consider for example a selective deletion of certain entries in a database. Once the place in the memory holding some data is overwritten, the original data are deleted. For example, the string "birthday happy" may be deleted from the database and be overwritten by "happy birthday". This is a typical scenario in classical computing systems. But it differs from the case of information dissemination within the computing system, say, when parts of the computer's memory are compressed and copied from one register to another, decreasing the system's AI complexity over time by means of self-organisation through data compression and the structuring of unstructured data.

However, strictly speaking, an IP account based on AI will have a limited capacity to explain cases in which some data are deleted and/or modified whilst the overall AI complexity does not decrease. Consider a system that starts with an input string S_{INPUT} encoding less than or equal to the information encoded by the output string S_{OUTPUT}. Unless the particular UTM or, say, a combination of finite state transducers running the program is *changed* as a result of the deletion of information, AIT cannot account for the deletion operation.

It seems then that, as expected, the processing of AI cannot be easily decoupled from the underlying computing system running either the algorithm or program. The underlying architecture of the computing system and the supported instructions are implicitly included in the calculation of AI complexity. Both conventional AIT and FSTAIT deal with idealised computation. They deal with what happens between input and output whilst *assuming* faultless computation (Calude, personal communication). Any possible errors during the actual computation are ignored. AIT is based on idealised UTMs and FSTAIT is based on faultless transducers. The processing of AI certainly entails computation, simply because AI is *defined* in terms of some computing system. But as an account of computation, an AI-based IP account is inadequate.

6.5.2 *Problems for an IP Account Based on Factual Information*

Arguably, information construed as *factual* information can lead to knowledge-that (Dretske, 1981, pp. 45–47; Floridi, 2008, p. 118). To reprise, the age-old debate concerning information and knowledge was one of the reasons for the objection discussed in Section 6.3.1. An IP account based on factual information has the implication that the information produced by digital computing systems *can* yield knowledge-that. This *can modality* should not be taken lightly. Such knowledge is either derived by its user (or programmer) or intrinsic to the computing system. The former option may seem at least at first blush unproblematic. Accordingly, any knowledge yielded is *derivative* and used by an external knower that interprets the information, which was produced by the computing system. The latter option seems more problematic, provided, of course, that knowledge *is*, and *not just* can be, yielded by the computing system in the absence of an external knower.

To see why such can modality is problematic, let us make some brief observations about *knowledge*. The Platonist definition of knowledge as a justified true belief (JTB) has been widely accepted in modern philosophy.[21] More recently, some have argued that understanding of knowledge in terms of information rather than beliefs is preferable (Dretske, 1981; Floridi, 2011, pp. 211–212). Floridi has argued that the idea of knowledge as JTB should be supplanted by knowledge as *correctly accounted for* factual information. The very idea of analysing knowledge on a doxastic basis in terms of JTB is supposedly misguided, since Gettier-type counterexamples are unavoidable *in principle* no matter how the JTB account is revised. But this simply shows that much of the burden of *knowledge* is shifted to *information*. Factual information has to tell us something *true* about a particular object or state of affairs. Depending on what qualifies as justification or evidence for the factual information concerned to upgrade to knowledge, even derivative knowledge, which may be extracted from computing systems, is highly questionable.

The other option of computing systems having the capacity of yielding intrinsic knowledge has been challenged by many philosophers (Agassi, 1988, 2003, pp. 601–603; Dretske, 1993; Dreyfus, 1979; Harnad, 1994; Penrose, 1989, pp. 531–532; Searle, 1980). It is not at all clear that there is compelling evidence to support such a capacity. There is only a limited sense in which a digital computing system "understands" or "knows" what some processed data may represent. A digital computer only understands specific machine instructions well enough to execute them. As already observed in Chapter 1, the "know-how" of the CPU, for example, requires no "know-that" of, say, the primitive ADD operation that the CPU can perform.

The semantics of machine instructions can be traced back from the higher level programming language, provided that the computing system is program-controlled, through assembly language to the physical operations of the logic circuits. The semantics of programming languages is formal and describes the relation between symbols of the programming language and their specific machine implementation. This formal semantics provides an abstract definition of the internal state of the computer and interprets the primitives of the programming language as actions on this state. A high level language, such as C++ or Java, describes the computer's state at a high level of abstraction referring to data structures and operations on them. A low level language, such as C or assembler, describes labelled relative memory addresses and operations on them (White, 2011, p. 194).

Moreover, as mentioned in Chapter 1, any factual information entered as input at the program level, is converted into something recognisable by the computing system by using an implicit semantics dictionary. At the hardware level, the working of the computing system is purely physical and is governed by laws of physics. Still, whatever goes on at this level is completely determined by the programmed instructions, the state of the computing system and the input entered (subject to the presence of noise).

---

[21] Edmund Gettier has famously challenged this definition arguing that truth, belief, and justification are not sufficient conditions for knowledge (1963). Gettier showed that a true belief might be justified and nevertheless fail to be knowledge.

Be that as it may, it does not follow that the computer manifests any *beliefs* that are associated with the operations it performs. Suppose that a doorbell is replaced with a digital computing system that emits the sounds: "someone is at the door", only when someone pushes the door button. When someone pushes the button, the system picks up the information about it, processes it and emits the sounds above as output. However, this output is not a *belief* of that system that someone *is* at the door, anymore than the doorbell would have believed that (Dretske, 1981, p. 204). One may further distinguish between information conveyed by assertions and displays (Sorensen, 2007, pp. 158–179). A computer program that *displays* information about rain in London tomorrow, it does not *assert* that it will rain in London tomorrow, even though this output may be based on a reliable source of information. There is no relevant belief inherent in such computing systems.

Even if one endorses a factual information-based, rather than a JTB-based, account of knowledge, the processing of data by a computing system proceeds regardless of the veridicality of the underlying data. Gricean non-natural meaning of signs (e.g., propping up a car's hood/bonnet indicating that the car is broken down) does not require a correspondence to the state of affairs in question (e.g., whether the car is actually broken down). Likewise, the information processed by the computing system can, but need not, correspond to an external state of affairs. Even if we took the input and the initial state as an external state of affairs (in the sense that they are set from outside the computing system), this would be the "point of departure" for the computational process itself. The process would proceed without necessarily preserving any correspondence between the processed information and the relevant external state of affairs. Whether the resulting processed information consistently represents some state of affairs or not is a contingent fact. We conclude that deterministic computation need not process factual information.[22] Even *if* some computational artificial agent were designed specifically to process factual information, this would merely show that some computation could process factual information.

6.6 The Instructional Information Processing (IIP) Account

It should come as no surprise then that one conclusion of the present chapter is that only the processing of instructional information entails concrete digital computation. And as already suggested in the analysis of the requirements of physical system to compute, it is the processing of instructional information that distinguishes trivial from nontrivial computing systems. The former are systems that process *data* that may be either structured or not but need not rise to instructional information, whereas the latter systems require the processing of instructional information. Let us see why.

---

[22] It is similarly argued in Piccinini and Scarantino (2011, p. 33) that digital computation does not entail the processing of truth-entailing semantic information about the distal environment.

6.6.1 An Outline of the IIP Account

According to the IIP account, nontrivial digital computation is the processing of digital data in accordance with finite instructional information.[23] A trivial computing system has only one capacity, and it, therefore, need not process instructional information. A nontrivial computing system has at least two capacities and, therefore, in general, it needs to process instructional information. A capacity is an action a computing system, whether abstract, such as an FSA, or physical, such as an iPad, *can* reliably execute. Each transition specified in the TM's machine table, for example, represents a capacity of the TM to move between successive complete configurations. An instruction is needed for executing the capacity of the system.

For all genuine computational systems (trivial or not) – what matters to their data processing is the abstract form (i.e., the *type*) of the data, rather than the particular implementation of the (i.e., *token*) data. The same data processing may take place in a variety of physical substrates. That is the root of the multiple realisability of algorithms and data. Multiple realisability of digital data is reflected in the definition of data as lack of uniformity. This definition leaves the matter of the implementation of digital data underdetermined (i.e., data are ontologically neutral (Floridi, 2011, p. 86)).

The processing of digital data in accordance with finite instructional information is clearly compatible with the realisation of effective procedures. Unconditional instructional information is the basis for the default control structure of sequential processing of instructions. Conditional instructional information is the basis of conditional branching. It allows the program to follow alternative paths of execution by using *If X Then Y Else Z* structures (where Y and Z are instructions). Similarly, the combination of these two forms of instructional information is the basis for the iterative control structure. A loop construct is used to repeat a set of instructions as long as a certain condition is met and a recursive method uses an explicit information trail to track the recursive case splits.

Whilst instructional information is clearly compatible with the imperative programming paradigm, it is also compatible with the functional and logical paradigms.[24] Programs belonging to the imperative (or procedural) type are characterised as *stateful*.[25] An imperative-type program (written, say, in C, Java or Pascal) is a sequence of instructions, which change the program's state as well as

---

[23] This account is significantly elaborated and formulated more precisely in Fresco & Wolf (forthcoming).

[24] The classification of all programs into these three paradigms is by no means exclusive or exhaustive. In some respects, the functional and the logical paradigms overlap. Prolog, for example, is commonly classified as either a functional or a logical programming language. Also, a functional programming language may have some imperative aspects, for instance.

[25] 'Stateful' means that a program keeps track of the state of interaction among subroutines, threads or processes, typically by setting values in shared memory. 'Stateless' means that there is no record of previous interactions and each interaction request is handled based only on data associated with it.

the state of the implementing computing system. A functional-type program (written, say, in PCF or Lisp) is a set of function definitions, which specify rules for operating on data to transform them into other data. The evaluation of these functions determines the resulting state of the *implementing* computing system. A logic-type program (written in, say, Prolog) is a set of rules and assertions from which a proof is constructed as a solution to a problem. Unlike imperative-type programs, programs belonging to either the functional or logic types are characterised as *stateless*. However, both functional-type and logic-type programs make use of function nesting (i.e., functions that are passed as arguments to other functions). The nesting of functions is underpinned by rules that operate on data, which are transformed into other data, in a manner that is likewise compatible with instructional information.

As remarked in Section 6.4, instructional information and control information are equivalent. The behaviour of a TM is controlled by its finite-state controller. At each step of the computation, the combination of the symbol read from the tape (i.e., the data) and the state of the TM uniquely determines the capacity that is exercised. The control information of special-purpose TMs is fixed in their internal descriptions and machine tables. Whilst a UTM actualises control information in the same way as the simulated special-purpose TM, it also actualises control information in a different way, giving rise to its universality. The UTM actualises control information that is encoded on its tape as input as part of the description of the simulated special-purpose TM (Wolf, 2013). It is worth noting that the actualisation of control information so-construed does not yield any "knowledge-that" of the information processed, just possibly "knowledge-how".

The behaviour of nontrivial computing systems is controlled by instructional information, since such systems have at least two alternatives of possible action. Any system that is capable of more than one action needs to include at least two control sub-states that can yield different actions. The selection between the different control sub-states is enabled by instructional information. The behviour of trivial computing systems, on the other hand, does not require instructional information. Trivial computing systems may be able to send, receive and transform data, which need not be structured. They transform data in a manner that, as in the case of nontrivial computation, depends on the type of the data. This transformation may be either a nonrandom transformation of one type of data to another or a nonrandom modification of data. A NOT gate is an example of a trivial computing system of the former type. If it gets data type 0 as input, it converts that 0 into a 1 and sends it as output. An AND gate is an example of the latter type. If it receives either (0, 1) or (1, 0) as input, it modifies the input into a 0 and sends it as output. In both cases, the gate exercises a *single capacity*.

The principle of actualising control information also applies to some discrete connectionist networks. Some connectionist networks only qualify as trivial computing systems. Consider a simple connectionist network for computing logical disjunction that consists of two input neurones and a single output neurone. It does not require any instructional information for its computation. The weighted edges between each of the two input neurones and the output neurone are of equal value, since the network should respond in the same manner for both (0, 1) and

(1, 0) inputs. No instructional information is required to differentiate the behaviour of this network for different inputs. On the other hand, connectionist networks that are capable of performing multiple operations (say, add and subtract) require instructional information for selecting among the different operations, since they have more than one capacity.

6.6.2 The IIP Account Evaluated

We now turn to evaluate the IIP account against the six recommended adequacy criteria. First, the IIP account meets the conceptual criterion, for it can explain the core concepts of digital computation. Although 'algorithm' is hard to rigorously define, it naturally lends itself to an explanation in terms of instructional information. Turing characterised 'algorithm' as an effective sequence of instructions (1936). In the same spirit, Kleene defined 'algorithm' as a method given by a finite set of rules or instructions that specifies a decision procedure (Shoenfield & Kleene, 1995, pp. 15–17). Arguably, an algorithm cannot be rigorously defined in full generality, for the extension of this concept is still expanding. In addition to sequential algorithms (which are the basis of Turing's characterisation), this concept has expanded to also refer to parallel, interactive, real-time and even quantum algorithms. Still, according to Yuri Gurevich, deterministic sequential algorithms can be defined by means of formal axiomatisation (2012). Informally, the first proposed axiom states that "[a]n algorithm determines a sequence of 'computational' states for each valid input" (Dershowitz & Gurevich, 2008, p. 306). The third axiom states that "[t]he transitions from state to state in computational sequences are *governable by some fixed, finite description*" (ibid, italics added).

Importantly, for our purposes, it does not matter so much whether a rigorous definition of an algorithm is feasible or not. What seems to underlie all the standard definitions of 'algorithm' is that "[it] is given as *a set of instructions* of finite size" (Dershowitz & Gurevich, 2008, p. 310, italics added). It follows then that instructional information accurately reflects the very essence of an algorithm. *Unconditional* instructional information underpins the default control structure in which instructions are followed sequentially and *conditional* instructional information underpins conditional branching control structure. These two forms of instructional information suffice to describe any algorithmic solution of a given Turing-computable problem.

Similarly, a program can also be explained in terms of instructional information. Programs are intended to produce some behaviour by executing them on some physical computing system either directly or indirectly. Instructional information and discrete data suffice to explain the behaviour of any computer program in any language. As shown by Corrado Böhm and Giuseppe Jacopini, any computer program whatsoever can be rewritten using just three control structures: sequences (i.e., do this; then do that), alternative selection (i.e., conditionals) and iteration (or looping) (1966). As discussed above, these three control structures can be described as instances of instructional information.

Since the concepts of algorithms and programs are explicable by the IIP account, compilation and interpretation may be simply reduced to (perhaps meta-) explanations of programs. Typically, computer programs are written in some high level programming language that the CPU cannot execute directly. To allow the source code of the program to be executed it is first converted into machine code. This conversion is accomplished by means of compilation or interpretation (depending on the type of programming language). Whereas a compiler only performs this conversion once, an interpreter usually converts it "on the fly" every time a program (or a single instruction) is executed. Both the compiler and the interpreter can thus be viewed as meta-programs that enable the source program to be converted into the target program, which can then be executed by the machine.

Second, the IIP account meets the dichotomy criterion. Paradigmatic cases of digital computing systems, such as CPUs, iPads, conventional digital computers, calculators, TMs and certain discrete connectionist networks are deemed computational on this account. For they all process discrete data in accordance with finite instructional information. The control unit in the CPU receives the part of the instruction that encodes a command (such as addition). This command, being the instructional information, is used by the controller to determine which operation needs to be performed on the corresponding data that are subsequently sent to another component for execution. Likewise, in a TM, the combination of the symbol read from the tape and the state of the TM determines the capacity to be exercised next. The same analysis applies even to the (now outdated) punch-card computers, in which programs were entered in the form of a sequence of punch cards.

Let us examine the case of nonprogrammable calculators. Calculators accept two types of input from their environment, namely discrete data and a specific instruction. The instruction causes the calculator's processing unit to perform a specific operation on the data by means of transforming them into the appropriate results. The memory units of the calculator hold the data (and possibly an intermediate result) until the operation is performed. The processing unit is typically limited to performing only one operation on the data at any given time. The particular operation of the calculator is determined uniquely by the instruction that was entered as input (possibly yielding an error message for illegal input data) (Piccinini, 2008, p. 35). It can easily be seen that calculators process discrete data in accordance with finite instructional information and are, thus, classified as nontrivial computing systems on the IIP account.

Is the IIP account so broad that it also classifies non-computing systems as computational? It is easy to argue that it is not too broad. The IIP account is certainly broad enough to encompass non-symbolic computing systems, such as discrete connectionist networks and logic circuits, as observed above. A simple microphone, on the other hand, is excluded by the IIP account as non-computational despite its capacity to receive and send information. For it fails to meet other key requirements: it does not store information, nor process information nor has the capacity to actualise control information. Similarly, planets do not compute on the IIP account. They move along ellipses revolving around a star in accordance with Kepler's laws of motion. Yet, such movement does not amount to planets processing instructional information. Any computational description applied to the motion of planets may only be applied *ex post facto* (Fresco & Wolf, forthcoming).

Third, the IIP account meets the taxonomy criterion too. Two-input, one-output logic gates, such as AND, OR and XOR gates, compute only trivially, because they have a single capacity that does not require instructional information to be exercised. They are less computationally powerful than complex logic circuits, such as half adders and full adders, for the above logic gates process unstructured data, whereas half adders and full adders process structured data, which need not rise to instructional information. Flip-flops and shift registers are memory-based and have more than one capacity thereby requiring instructional information for their computation. Special-purpose digital computers are more computationally powerful than flip-flops and shift registers, for the former have the requisite capacities to compute, in principle, any *particular* Turing-computable function. Depending on their particular configurations, special-purpose digital computers also vary in their computational power. Lastly, by virtue of their (approximate) universality, programmable digital computers are computationally more powerful than any special-purpose digital computer. For they can store multiple programs and thereby process an arbitrarily large number of instructions that enable the requisite capacities.

Fourth, the IIP account also meets the implementation criterion, for it is not limited to the program or algorithmic level. The computer's architecture can be interpreted as a relatively neutral medium, e.g., as an array of bytes, that supports a wide range of information processes (MacLennan, 2011, pp. 235–236). But even the hardware level, which is traditionally viewed as a single level, can be further analysed into several sublevels[26] that are explicable in information processing terms. At the *functional* level – the operation of the computing system is analysed in terms of the function being computed in the process of the underlying registers changing their stored values. At this level of abstraction, the IIP analysis is applicable in a like manner to the algorithmic level. At the *register transfer* level – the operation of the computing system is analysed in terms of registers changing stored discrete data in accordance with instructional information. At the *logical* level – the operation of the computing system is analysed in terms of logic gates operating on digital data.

Fifth, the IIP account meets the miscomputation criterion as well. Different types of miscomputation are explicable in terms of instructional information processing. Hardware malfunctions (e.g., a hard-drive malfunction), for example, can be explained as *noise* on the communication channel that results in either corrupted data, which cannot be further processed, or the failure to send and receive data for processing. Other errors resulting from, say, an illegal input, can

---

[26] This hierarchical decomposition is also consistent with Newell's analysis of the standard architecture of a conventional digital computer (Newell, 1980, p. 173).

be explained as either the *lack* of expected data or the data not being *well-formed* (cf. the first and second requirements of SDI, respectively) that are required as input for the computational process to resume. A trivial computing system, such as an OR gate, is said to miscompute when in the presence of noise it exercises a capacity it does not otherwise have.

Finally, it is self-evident that the IIP account meets the program execution criterion too. Some accounts *identify* digital computation with the execution of programs (cf. the PSS account and the algorithm execution account). According to the IIP account, every digital system that executes programs is genuinely computational, but the converse does not hold. Some digital systems, such as flip-flops and discrete connectionist networks, do not execute programs and yet, they are classified as computational.

6.7 Concluding Remarks

Importantly, while strictly SI-based or AI-based IP accounts are inadequate for explaining concrete computation, SI and AI are systematically related to instructional information. The strong relation between SI and AI is evident. For instance, strings with low AI complexity are less informative in Shannon's sense. Random strings have high AI complexity are more informative in Shannon's sense (Adriaans & van Benthem, 2008, p. 6). Moreover, as observed above, both SI and AI can be used to measure thermodynamic entropy in computing systems that process information.

But there are also systematic relations between SI and instructional information as well as between AI and instructional information. SI can be instrumental for the analysis of information flow in computer programs (cf. Malacaria, 2007). On the one hand, the program's states, relevant variables and loops or conditionals can be analysed in terms of data and instructional information. On the other hand, the program can be treated as a black box and its output, which is visible to a potential attacker, can be analysed in terms of SI to quantify information leakage, for example. AIT is inherently based on the *qualitative* notion of instructional information. AI complexity essentially depends on the number of instructions the program performs on a given input, the computational constructs of the particular programming language used (in terms of the basic instructions that are supported) and the size of the program in bits (Calude, 1988, p. 383).

As we have seen, the IIP account is adequate for the explanation of concrete digital computation. First, it avoids the problems faced by the resulting accounts that are based on SI, AI and factual information. Second, it meets all the six criteria for evaluating the adequacy of accounts of concrete digital computation. The relations among instructional information, AI and SI call for further analysis in the context of computation and thermodynamics, given that information processing in the course of physical computation involves the dissipation of varying amounts of energy.

References

Ackoff, R.: From Data to Wisdom. Journal of Applied Systems Analysis 16, 3–9 (1989)

Adriaans, P.: Learning and the cooperative computational universe. In: Adriaans, P., van Benthem, J. (eds.) Philosophy of Information, vol. 8, pp. 133–167. Elsevier, Amsterdam (2008)

Adriaans, P.: A Critical Analysis of Floridi's Theory of Semantic Information. Knowledge, Technology & Policy 23(1-2), 41–56 (2010), doi:10.1007/s12130-010-9097-5

Adriaans, P.: Information. In: Zalta, E.N. (ed.) The Stanford Encyclopedia of Philosophy (2012a), http://plato.stanford.edu/archives/win2012/entries/information/ (retrieved)

Adriaans, P.: Facticity as the amount of self-descriptive information in a data set. CoRR, abs/1203.2245 (2012b), http://arxiv.org/pdf/1203.2245v1 (retrieved)

Adriaans, P., van Benthem, J.: Introduction: Information is What Information Does. In: Philosophy of Information, vol. 8, pp. 3–26. Elsevier, Amsterdam (2008)

Adriaans, P., van Emde Boas, P.: Computation, Information, and the Arrow of Time. In: Computability in Context, pp. 1–17. Imperial College Press (2011)

Agassi, J.: Winter 1988 Daedalus. ACM SIGART Bulletin (105), 15–22 (1988), doi:10.1145/49093.1058127

Agassi, J.: Newell's list. Behavioral and Brain Sciences 26(05) (2003), doi:10.1017/S0140525X03220136

Baez, J.C., Stay, M.: Algorithmic Thermodynamics. arXiv:1010 (2010)

Bais, F.A., Farmer, J.D.: The Physics of Information. In: Adriaans, P., van Benthem, J. (eds.) Philosophy of Information, vol. 8, pp. 609–683. Elsevier, Amsterdam (2008)

Bar-Hillel, Y., Carnap, R.: An Outline of a Theory of Semantic Information (Technical Report No. 247), pp. 1–49. Research Laboratory of Electronics, MIT (1952)

Bechtel, W.: Representations and cognitive explanations: Assessing the dynamicist's challenge in cognitive science. Cognitive Science 22(3), 295–317 (1998), doi:10.1016/S0364-0213(99)80042-1

Bennett, C.H.: Logical Reversibility of Computation. IBM Journal of Research and Development 17(6), 525–532 (1973), doi:10.1147/rd.176.0525

Böhm, C., Jacopini, G.: Flow diagrams, turing machines and languages with only two formation rules. Communications of the ACM 9(5), 366–371 (1966), doi:10.1145/355592.365646

Bohm, D.: The special theory of relativity. Routledge, London (2006)

Burgin, M.: Theory of information: fundamentality, diversity and unification. World Scientific Pub. Co. Inc., Hackensack (2010)

Calude, C.S.: Theories of computational complexity. Elsevier Science, New York (1988)

Calude, C.S.: Information and randomness: an algorithmic perspective (2nd ed., rev. and extended.), New York (2002)

Calude, C.S.: Information: The Algorithmic Paradigm. In: Sommaruga, G. (ed.) Formal Theories of Information. LNCS, vol. 5363, pp. 79–94. Springer, Heidelberg (2009)

Calude, C.S., Salomaa, K., Roblot, T.K.: Finite state complexity. Theoretical Computer Science 412(41), 5668–5677 (2011), doi:10.1016/j.tcs.2011.06.021

Chaitin, G.J.: Information, randomness & incompleteness: papers on algorithmic information theory, 2nd edn. World Scientific, Singapore (1990)

Chaitin, G.J.: Algorithmic information theory. Cambridge University Press, Cambridge (2004)

Cleveland, H.: Information as a resource. The Futurist 16(6), 34–39 (1982)

Cordeschi, R.: Cybernetics. In: Floridi, L. (ed.) The Blackwell Guide to the Philosophy of Computing and Information, pp. 186–196. Blackwell, Malden (2004)

Corning, P.A.: "Control information": The missing element in Norbert Wiener's cybernetic paradigm? Kybernetes 30(9/10), 1272–1288 (2001), doi:10.1108/EUM0000000006552

Dershowitz, N., Gurevich, Y.: A natural axiomatization of computability and proof of Church's Thesis. Bulletin of Symbolic Logic 14(3), 299–350 (2008), doi:10.2178/bsl/1231081370

Downey, R.G., Hirschfeldt, D.R.: Algorithmic randomness and complexity. Springer, New York (2010)

Dretske, F.I.: Knowledge & the flow of information. MIT Press, Cambridge (1981)

Dretske, F.I.: Can intelligence be artificial? Philosophical Studies 71(2), 201–216 (1993), doi:10.1007/BF00989857

Dreyfus, H.L.: What computers can't do: the limits of artificial intelligence. Harper & Row, New York (1979)

Fetzer, J.H.: Information: Does it Have To Be True? Minds and Machines 14(2), 223–229 (2004), doi:10.1023/B:MIND.0000021682.61365.56

Feynman, R.P.: Feynman lectures on computation. Addison-Wesley, Reading (1996)

Floridi, L.: Is Semantic Information Meaningful Data? Philosophy and Phenomenological Research 70(2), 351–370 (2005), doi:10.1111/j.1933-1592.2005.tb00531.x

Floridi, L.: Trends in the Philosophy of Information. In: Adriaans, P., van Benthem, J. (eds.) Philosophy of Information, vol. 8, pp. 113–131. Elsevier, Amsterdam (2008)

Floridi, L.: Information: a very short introduction. Oxford University Press, Oxford (2010)

Floridi, L.: The philosophy of information. Oxford University Press, Oxford (2011)

Fresco, N., Wolf, M.J.: The instructional information processing account of digital computation. Synthese (forthcoming), doi:10.1007/s11229-013-0338-5

Gandy, R.: Church's Thesis and Principles for Mechanisms. In: The Kleene Symposium, pp. 123–148. North-Holland (1980)

Gettier, E.L.: Is Justified True Belief Knowledge? Analysis 23(6), 121–123 (1963), doi:10.2307/3326922

Green, M.: Imperative logic. In: Craig, E. (ed.) Routledge Encyclopedia of Philosophy. Routledge, London (1998), http://www.rep.routledge.com/article/X043 (retrieved)

Gurevich, Y.: What Is an Algorithm? In: Bieliková, M., Friedrich, G., Gottlob, G., Katzenbeisser, S., Turán, G. (eds.) SOFSEM 2012. LNCS, vol. 7147, pp. 31–42. Springer, Heidelberg (2012)

Hamblin, C.L.: Imperatives. Basil Blackwell, New York (1987)

Harnad, S.: Computation is just interpretable symbol manipulation; cognition isn't. Minds and Machines 4(4), 379–390 (1994), doi:10.1007/BF00974165

Hintikka, J.: Some varieties of information. Information Processing & Management 20(1-2), 175–181 (1984), doi:10.1016/0306-4573(84)90047-5

Landauer, R.: Irreversibility and Heat Generation in the Computing Process. IBM Journal of Research and Development 5(3), 183–191 (1961), doi:10.1147/rd.53.0183

Li, M., Vitányi, P.M.B.: An introduction to Kolmogorov complexity and its applications. Springer, New York (2008)

Lucey, T.: Management information systems. DP Publications, Eastleigh (1987)

MacLennan, B.J.: Bodies — both informed and transformed embodied computation and information processing. In: Dodig-Crnkovic, G., Burgin, M. (eds.) Information and Computation, pp. 225–253. World Scientific (2011)

Malacaria, P.: Assessing Security Threat of Looping Constructs. In: Proceedings of the 34th ACM Symposium on Principles of Programming Languages, pp. 225–235 (2007)

Newell, A.: Physical Symbol Systems. Cognitive Science 4(2), 135–183 (1980), doi:10.1207/s15516709cog0402_2

Okun, L.B.: Energy and mass in relativity theory. World Scientific, Singapore (2009), `http://public.eblib.com/EBLPublic/PublicView.do?ptiID=477224` (retrieved)

Penrose, R.: The emperor's new mind: concerning computers, minds, and the laws of physics. Oxford University Press, Oxford (1989)

Piccinini, G.: Computing mechanisms. Philosophy of Science 74(4), 501–526 (2007), doi:10.1086/522851

Piccinini, G.: Computers. Pacific Philosophical Quarterly 89(1), 32–73 (2008), doi:10.1111/j.1468-0114.2008.00309.x

Piccinini, G., Scarantino, A.: Information processing, computation, and cognition. Journal of Biological Physics 37(1), 1–38 (2011)

Popper, K.R.: The Logic Of Scientific Discovery (14th Printing). Routledge, London (2002)

Ralston, A.: Encyclopedia of computer science, 4th edn. Nature Pub. Group, New York (2000); Reilly, E.D., Ralston, A., Hemmendinger, D. (eds.)

Scarantino, A., Piccinini, G.: Information without truth. Metaphilosophy 41(3), 313–330 (2010), doi:10.1111/j.1467-9973.2010.01632.x

Searle, J.R.: Minds, brains, and programs. Behavioral and Brain Sciences 3(03), 417–424 (1980), doi:10.1017/S0140525X00005756

Shannon, C.E.: A mathematical theory of communication. Bell System Technical Journal 27, 379–423, 623–656 (1948), doi:10.1145/584091.584093

Shoenfield, J.R., Kleene, S.C.: The Mathematical Work of S. C. Kleene. The Bulletin of Symbolic Logic 1(1), 9–43 (1995), doi:10.2307/420945

Sloman, A.: What's Information, For An Organism Or Intelligent Machine? How Can A Machine Or Organism Mean? In: Dodig-Crnkovic, G., Burgin, M. (eds.) Information and Computation, pp. 393–438. World Scientific (2011)

Sorensen, R.: Can the dead speak? In: Nuccetelli, S., Seay, G. (eds.) Themes from G.E. Moore New Essays in Epistemology and Ethics, pp. 158–180. Oxford University Press, Oxford (2007)

Teixeira, A., Matos, A., Souto, A., Antunes, L.: Entropy Measures vs. Kolmogorov Complexity. Entropy 13(12), 595–611 (2011), doi:10.3390/e13030595

Turing, A.M.: On Computable Numbers, with an Application to the Entscheidungsproblem. Proceedings of the London Mathematical Society s2-42(1), 230–265 (1936), doi:10.1112/plms/s2-42.1.230

White, G.: Descartes Among the Robots. Minds and Machines 21(2), 179–202 (2011), doi:10.1007/s11023-011-9232-4

Wiener, N.: Cybernetics: or, Control and communication in the animal and the machine. MIT Press, Cambridge (1948)

Wiener, N.: God and Golem, Inc.: a comment on certain points where cybernetics impinges on religion. MIT Press, Cambridge (1966)

Wolf, M.J.: The Importance of Actualizing Control in the Processing of Instructional Information. Philosophy & Technology 26(1), 67–70 (2013), doi:10.1007/s13347-012-0076-5

Chapter 7
Causal and Functional Accounts of Computation Examined

In this chapter, we examine three fairly recent accounts of computation with at least one common characteristic, namely, that they do not posit any extrinsic representational properties. These accounts imply that concrete digital computing systems can be individuated by either their causal properties or their functional/organisational properties. Whilst the third account examined in this chapter is also causal in essence, it emphasises the importance of the relevant *functional*/organisational properties of digital computing systems in addition to their causal properties.

The three accounts of concrete computation discussed here are the Gandy-Sieg account, the algorithm execution account and the mechanistic account. The first two accounts belong to the causal view of computation and the third one to the functional view of computation. According to the Gandy-Sieg account, a physical system computes if it goes through a sequence of state transitions whose input is encoded as the system's initial state, and each one of its states is its output at a given time (Gandy, 1980, p. 127). According to the algorithm execution account, a physical system computes if it acts in accordance with an algorithm (Copeland, 1996, 1997, p. 696). According to the mechanistic account, a physical system computes if it manipulates input strings of digits, depending on the digits' type and their location in the string, in accordance with a rule defined over these strings and possibly the system's internal states (Piccinini & Scarantino, 2011, p. 8).

7.1 The Gandy-Sieg Account

7.1.1 Introduction

According to Gandy, concrete digital computation amounts to a sequence of state transitions of a *Gandy machine* (GM), whose input is encoded as the machine's initial state, and each one of the machine's, possibly infinite, states is its output at a given time (1980, p. 127). Wilfried Sieg's analysis of computation subsumes Gandy's analysis, but offers a simplified version by excluding many technical details, which are present in the latter (Sieg & Byrnes, 1999; Sieg, 2007). Since our focus here is on the conceptual framework of GM computation, we can, for the most part, set aside the mathematical formulations that appear in Gandy's original analysis and consider the Gandy-Sieg account instead. To do justice to this account as an explanation of concrete digital computation it should also be examined in comparison with Turing's account.

Whereas Turing's original analysis was motivated by human calculability, Gandy's aim was to formulate a general notion of *machine* computation. Turing's analysis can be applied to *either* humans or machines, but some crucial aspects of his analysis rely on a calculation carried out by a human (Sieg & Byrnes, 1999, p. 150). For example, human calculation proceeds as a sequence of elementary steps, whereas an *artificial* physical machine can process an arbitrarily large number of symbols *concurrently* (Gandy, 1980, pp. 124–125). By violating the boundedness condition imposed by Turing's account, the Gandy-Sieg account extends the former to include parallel computation. This is the essential difference between these two accounts as is shown below.

Moreover, Turing's analysis of computable functions by idealised TMs may arguably be inapplicable to some imaginable computing systems. These systems crucially include John Conway's "Game of Life" and cellular automata that perform computation in parallel (Sieg & Byrnes, 1999, p. 150).[1] Turing's account assumes that "the computation is carried out on one-dimensional paper, i.e. on a tape divided into squares" (1936, p. 249). In a manner akin to human calculations on symbol sequences, Turing's account relies on a *low*-dimensional symbolic configuration. Yet, according to Gandy, machines can in principle carry out parallel computations as evidenced by, say, cellular automata operating on *multi-dimensional* configurations.

A GM is a discrete deterministic system, since Gandy explicitly excluded analogue machines from consideration. By his lights, the system under consideration is, in a loose sense, a *digital computer* (Gandy, 1980, p. 126). Such a machine can be mathematically described as the pair (S, F), where S is a set of the system's states and F is the state transition function over S. Operations on these states are composed of local transformations that modify "parts" of a bounded size resulting in configurations that are uniquely assembled, up to isomorphism, into the next state of the machine. The basic idea is that a GM has to recognise the causal neighbourhoods of a given state from a bounded set in a given finite configuration, and act on them locally but concurrently and assemble the results of these local operations into the next configuration (Sieg & Byrnes, 1999, p. 151; Sieg, 2008, p. 146).

Whereas the states of a TM are modelled by means of strings of symbols, the states of a GM are represented by hereditarily finite sets. An hereditarily finite set is a set that can be constructed by using the pair formation operation {a, b} and

---

[1] It does not follow, of course, that TMs cannot compute functions that are computable by cellular automata. It was proven that any function computable by a cellular automaton is also computable by some TM.

the union operation on its members repeatedly, starting from the empty set $\{\}$.[2] The GM's states are built from a set of atoms that can be conceptualised as the fundamental components of the machine. Members of these hereditarily finite sets represent states of physical machines in the same manner that variables in physics represent states of nature (Sieg & Byrnes, 1999, pp. 152–153).

7.1.2 The Key Requirements Implied by the Gandy-Sieg Account

We now turn to examine the four requirements for a physical system to perform (parallel) digital computation, according to the Gandy-Sieg account.

- 1st Requirement: being amenable to a description in terms of a sequence of state transitions.
- 2nd Requirement: having a limited complexity of hierarchical structure.
- 3rd Requirement: having a unique assembly of finitely many elementary parts.
- 4th Requirement: being comprised of states that can only affect other states after some delay that is proportional to their spatial distance.

These requirements are in effect key constraining principles that Gandy introduced, arguably in consideration of *physical* machines. They are the basis for Gandy's Theorem stating that any function that can be calculated by a system satisfying them is Turing-computable (Gandy, 1980, p. 126).

According to the first requirement, the system should be describable, as remarked above, by the pair (S, F), where if $s_0 \in S$ is the system's initial state, then s_0, $F(s_0)$, $F(F(s_0))$ and so on are its subsequent states (Gandy, 1980, pp. 126–127). This description should reflect the actual, concrete structure of the system in a specific state (Sieg & Byrnes, 1999, p. 155). At the same time, it should be sufficiently abstract to also apply to mechanical, electrical or merely abstract systems, such as TMs. To that end, the various parts of the system, such as the teeth of cogwheels or the electrodes of a transistor, are labelled. Since Gandy's analysis applies to both abstract and concrete systems, labels are necessary as referring expressions for parts of the latter kind of systems. The labels are fixed such that, typically, if a label l refers to a specific part of a system in a particular state, then l refers to the *same* part in the next state too (Gandy, 1980, p. 128). Labels may also be used for positions in space, for example, the squares of the tape of a physically constructed TM, as well as for physical features of the system, such as the symbols on the tape of a physically constructed TM or the state of a transistor.

---

[2] The sets $\{\{\}\}$ and $\{\{\}, \{\{\}\}\}$, for example, are hereditarily finite and their "heights" are 1 and 2, respectively (a set's height is defined as the maximum depth of nesting brackets minus one). Also, all members of hereditarily finite sets are themselves hereditarily finite. For a simplified presentation of Gandy's analysis outlining relevant set-theoretic concepts, including hereditarily finite sets and transitive closure, see Sieg and Byrnes (1999).

According to the second requirement, each state $s_i \in S$ can be assembled from parts, which can be aggregates of other parts, and, yet, for any given system there is a finite bound on the complexity (the maximum height, in Gandy's terminology) of this hierarchical structure (Gandy, 1980, pp. 130–131). That is, the set theoretic rank of the states in S is bounded (Sieg & Byrnes, 1999, p. 155). Turing asserted that, at any point in time, the motion of the TM, the square just scanned, the complete sequence of all symbols on the tape and the m-configuration describe the total state of the TM (1936, p. 253). So if the human computor takes a break, this description of the total state allows her to resume the computing operation from that point. Similarly, for a GM, all the data about a particular state that are relevant to the operation of the system have to be encoded in any structure that is used to describe it (Gandy, 1980, p. 128).

Moreover, a system can be described in terms of hierarchies. For instance, a series of parallel computer processors may be subordinate to one main control unit. It is natural then when describing such a hierarchy to consider a higher-level part as having its subordinates as members in the hierarchical structure. This principle establishes that for any given computing system the maximum height of its hierarchical structure has to be finite.

According to the third requirement, each state $s_i \in S$ is uniquely assembled from elementary parts of a bounded size that are drawn from a repository of a bounded number of types of such elementary parts (Gandy, 1980, pp. 131–133). These parts can be labelled in a manner such that there is a unique way, up to isomorphism, of assembling them together. The state s_i following state s_j ($s_i, s_j \in S$) is determined uniquely up to an isomorphism over s_j (Sieg, 2008, p. 149). The tape of a TM, for example, can be uniquely reassembled from the collection of *all* pairs of consecutive squares with their symbols. If two such pairs have the same label for some square, then they are assembled together with an overlap.

According to the fourth requirement, for every state $s_i \in S$, the parts from which the next state, $F(s_i)$, can be assembled depend only on bounded parts of s_i (Gandy, 1980, pp. 135–136). In other words, each changed part of s_i is affected only by its locally bounded "neighbourhood" (Copeland & Shagrir, 2007, p. 219). This requirement was identified by Gandy as the most important one amongst all four. Turing's analysis required that an action of a TM depended only on a bounded portion of the tape. This requirement was based on a physical limitation of the modelled human computor. A TM can move to a different set of scanned squares, but only within a certain *bounded distance* of the scanned square. In Gandy's analysis this boundedness limitation is replaced by the physical limitation that causal neighbourhoods of s_i be structurally determined.

The fourth requirement is justified by the principle of finite velocity of propagation of effects and signals. For contemporary physics rejects the possibility of instantaneous action at a distance (Sieg & Byrnes, 1999, p. 158).[3]

---

[3] It is worth noting that the fifth requirement of the PSS account (where some actions of the physical computing process could also be "at a distance") is in tension with this physical principle.

One motivation for this requirement was that it should apply to both *abstract* and *concrete* systems, so Gandy's analysis did not distinguish spatial structures from non-spatial structures. Another motivation was the ability to determine the causal neighbourhoods of s_i without having an existing knowledge of the regions S of $F(s_i)$ (Gandy, 1980, p. 135). Given the limit on the velocity of signal propagation and assuming a homogenous structure of space and time, the state of the system in a particular region at a given time instant can be calculated based on a finite number of states in neighbouring regions at the previous time instant (Gandy, 1980, p. 136). Gandy took the limit on the velocity of signal propagation to be the speed of light and this was the subject of one of his critiques. For in Newtonian mechanics gravity is instantaneous and, hence, signals can propagate at infinite velocity (Dowek, 2012, pp. 28–29).

7.1.3 The Gandy-Sieg Account Evaluated

Each of the four requirements above is, arguably, necessary for deterministic digital computation. For instance, if the first requirement – the *form of description* – were violated, then the system would no longer be deterministic, since despite the availability of a complete description of its initial state, its subsequent behaviour would not be uniquely determined (Gandy, 1980, p. 126). As well, if the second requirement – the *limitation of hierarchy* – were violated and the complexity of the assembly of states were not finitely bounded, then the system would no longer be discrete. For Gandy's determinism presupposes that each state of the machine can be adequately described in *finite* terms (1980, p. 127).

Although these requirements seem restrictive, Gandy argued that they are not (1980, p. 145). He asserted that his analysis does not depend on favouring any particular set of basic operations. At each step only a bounded portion of the whole state is changed. A GM can perform the *basic* operations of most computational procedures in a single step. The first three requirements simply capture the discrete and deterministic nature of the GM. However, they do not explicitly specify that a halting computational process consist of *finitely many* steps. The fourth requirement above, *local causation*, restricts the type of operations available to the GM. Each changed part of a state is only affected by its local bounded neighbourhood (Shagrir, 2002, p. 233).

Whether this account meets the implementation criterion or not largely depends on the interpretation of a GM. There are two physical presuppositions, which are reflected in the four requirements above, that at least prima facie suggest that Gandy's original analysis was about concrete digital computation or physical machine computation. The first presupposition is the lower bound on the size of each atomic part of the GM. This lower bound makes the machine *discrete* and determines the maximal number of parts that can exist within a given neighbourhood. The second presupposition, which arguably applies to *all* physical systems, is that each atomic part is bounded in its ability to transmit and receive signals in its bounded neighbourhood (Sieg and Byrnes 1999: p. 162). The combination of these two presuppositions yields Turing's locality condition, that is, that the computing system can only directly affect finitely many different parts. But since parallel computation in physical systems is possible, the overall transition has to be uniquely determined by local actions on these bounded parts.

Nevertheless, the interpretation of a GM is ambiguous (Copeland & Shagrir, 2007; Shagrir, 2002). Shagrir offers three different interpretations, but argues that none of them provides evidence for claiming that Gandy's analysis pertains to finite *physical* computation (2002, p. 234). On one interpretation, Gandy's analysis applies only to a particular class of *physical* machines with the aim of proving that the functions computed by GMs are all Turing-computable. Gandy's analysis then applies to the computational bounds of the class of discrete deterministic computing systems that may be infinite (Shagrir, 2002, pp. 234–236).

However, this first interpretation is arguably problematic. For some cases of digital machines, such as the PMH machine, supposedly satisfy the four requirements above, but they exceed the class of Turing-computable functions. The PMH machine, conceived by Itamar Pitowsky, David Malament and Mark Hogarth, is a hypecomputer. That is, it can compute uncomputable or nonrecursive functions (Copeland, 2002, p. 461) by performing infinitely many computational steps in a finite time span in a kind of relativistic spacetime. The PMH machine is a discrete state machine consisting of two standard digital computers that communicate with each other. Arguably, it is deterministic, as its halting state is uniquely determined once a complete description of its initial state is given. It also satisfies Gandy's two physical presuppositions mentioned above. However, in cases where no signal is received from one of the two computers comprising the PMH machine, the first requirement above is violated (Copeland & Shagrir, 2007, pp. 226–228). If the Gandy-Sieg account is correct, "there can be no such thing as a discrete deterministic hypercomputer" (Copeland & Shagrir, 2007, p. 218), Nevertheless, the physical possibility of accelerating TMs in either an atomic universe or a quantum mechanical universe is questionable (Davies, 2001, pp. 677–679).

Two interpretations remain: the GM as either a purely abstract model of computation or a particular subclass of effective machine computation. According to the former, the GM is a mathematical notion, as is the TM, providing an abstract notation of *parallel* computation. The GM defines a machine mathematically while abstracting away from the properties of any particular class of computations by either physical or abstract finite machines. A physical system that does not satisfy the four requirements above, does not refute them, but simply falls outside their scope (Shagrir, 2002, pp. 235–236). This interpretation is supported by Sieg, who argues that "[t]he crucial difference between these two abstract models of computation [i.e., TMs and GMs] lies in the fact that [... GMs] operate on arbitrarily many bounded parts" (Sieg & Byrnes, 1999, p. 162).

But although Gandy extended Turing's analysis of algorithms to also incorporate parallelism, some algorithms such as the N-Queens problem, are excluded from Gandy's analysis. For instance, when the N-Queens problem is solved by a neural network through a finite number of state transitions[4], the fourth requirement above (i.e., local causation) may be violated. Each neurone from which a state $F(x)$ can be assembled depends on the values of the neurones of state x. Hence, if x and $F(x)$ are large enough, any neurone of $F(x)$ will depend on a part of x that exceeds the appropriate bound (Shagrir, 2002, pp. 228, 236).

It seems then that the least problematic interpretation of the GM is as effective *machine* computation (Shagrir, personal communication).[5] On this third interpretation, Gandy was set to characterise the *class* of algorithms, which are realisable by systems that comply with the four requirements above. This interpretation accords with Gandy's stipulation that he "shall distinguish between 'mechanical devices' and 'physical devices' and consider only the former" (1980, p. 126). Whereas Turing's analysis characterises the restrictive conditions on effective *human* computation, Gandy's analysis characterises the restrictive conditions on effective artificial machine computation. The latter recognises the possible variety of implementing computing systems and the fact that their physical resources are bounded (Gurevich, 1988). To be sure, it is not what functions are *computable* that is under consideration here, but rather *what* algorithms are *physically realisable*. If that is right, then the four requirements above simply apply to abstract parallel algorithms (Gurevich, 2012a).

Still, Gandy's analysis offers a more realistic approach to machine computation than Turing's analysis. For it considers the possible variety of implementing computing systems as well as the boundedness of their physical resources, including the limitation imposed by the velocity of signal propagation. In considering physical constraints, Gandy extended the *classical* notion of effective computability to apply to physical systems, which *could* in principle implement the algorithms in question. We may plausibly conclude upon examining the three possible interpretations above that the Gandy-Sieg account fails to meet the implementation criterion.

The Gandy-Sieg account seems to also fail to meet the dichotomy criterion. As Gurevich points out, very few modern real-world computing systems satisfy the four requirements above (2012a). One problem is the form of GM's states, which are defined as a collection of hereditarily finite sets. Any attempt to describe the state transitions of a nontrivial algorithm using hereditarily finite sets is very

---

[4] An initial state of the neural network consists of n^2 neurones, each representing the presence of a queen on a cell of the game board, and a matrix of n^4 weights between pairs of neurones.

[5] This interpretation reflects a shift from Shagrir's previous view of "Gandy machines as finite-physical machines" (Shagrir, 2002, p. 235). Yet, it is in line with his concluding remark that "Gandy's analysis applies *at best* to a proper subclass of finite computers, namely, to those satisfying [Gandy's four principles]" (ibid: p. 236, italics added). Shagrir prefers calling this interpretation an *intersection* between physical computation and abstract computation, inasmuch as the former is viewed as an implementation of algorithms. But this could raise some difficulties, if one views the implementation of an algorithm as a physical process that no longer has anything in common with an abstract algorithm.

difficult. Another problem is that GMs are synchronous parallel machines. For a group of autonomous processes to act in parallel as a *coherent* system, they have to be synchronised. Conway's game of life cellular automaton, for example, can grow without any bound *while preserving its synchronicity*. Yet, *real-world* cellular automata would not stay synchronous in a similar manner.

Even if we considered conventional digital computers, some of them would also be excluded by the Gandy-Sieg account, for they are asynchronous. For example, multi-core digital computers, say, those based on Intel core i5 or i7 technology, are typically asynchronous and can be used to implement cellular automata. Although a synchronisation of these cores *is* possible[6], at least in principle (Gurevich, personal communication), the Gandy-Sieg account excludes multi-core digital computers. Single core digital computers also exhibit high-level asynchrony despite the physical-level synchronicity. For example, they may run two different programs simultaneously. These programs are not synchronised, but the operating system allows them to run simultaneously by using time-sharing of the computational resources. The Gandy-Sieg account excludes such computing systems as well.

There are other computational models that are excluded by the Gandy-Sieg account. For example, a discrete neural network solving the N-Queens problem, which violates at least one of the four requirements above, is excluded even if it effectively computes a solution to the problem concerned. (However, to be sure, there exists a GM that does compute a solution to the N-Queens problem.) Additionally, the finiteness condition of Gandy's model excludes some useful algorithms, such as an algorithm for tracking the maximal numeral in a stream of numerals (Gurevich, 2012b). A TM can execute this algorithm, if the stream of numerals is inscribed on its initial tape, whereas a GM cannot.

In sum, although the Gandy-Sieg account offers a model of parallel synchronous computation, it is ultimately judged inadequate as an explanation of concrete digital computation. Turing's model was a human computor whose operations are as simple as they get and who has a single processor. Every operation of a TM only affects a square immediately to the right or left of the observed square. Gandy's model was not constrained by such human-oriented limitations. Whilst Gandy's analysis was motivated by physical limitations of deterministic discrete machines, important aspects of concrete computation were not addressed. For instance, his analysis remained silent about the reliability of computing systems and their susceptibility to malfunction and, hence, it also fails to meet the miscomputation criterion. The fourth requirement above is justified by the principle of *finite velocity* of signal propagation. However, there is no mention of the impact of noise on signal propagation and the underlying computation.

---

[6] A common clock can be used for all of the cores, in which case they would be synchronous.

7.2 The Algorithm Execution Account

7.2.1 Introduction

Computer science is typically defined as the science of algorithmic problem solving that studies the formal properties of algorithms and data structures as well as their mechanical and linguistic realisations (Gibbs & Tucker, 1986, p. 204). Yet, this characterisation crucially depends on what we take an algorithm to be. Whilst, informally, an algorithm is an ordered sequence of instructions that is guaranteed to solve a particular problem, a more precise characterisation of 'algorithm' is needed for a critical evaluation of the algorithm execution account of computation.

There are three main approaches to the characterisation of algorithms, according to Gurevich (2012a). The first one adopts a very abstract conception of 'algorithm' taking recursion as a primitive operation[7], but this approach is so broad as to also admit algorithms that, arguably, *cannot be physically implemented* (Moschovakis & Paschalis, 2008, p. 87), such as infinitary algorithms performing infinitely many steps before completing (Moschovakis, 1998, p. 88). The second approach, which is reflected in the Gandy-Sieg account, restricts the class of algorithms only to those, which are, at least, in principle physically realisable by discrete deterministic machines (Gurevich, 2012a, pp. 35–36). According to the third approach, an algorithm cannot be completely and rigorously defined; an analogy of algorithms to numbers suggests the reason for that.[8] There are many types of numbers, including positive integers, negative integers, naturals, rationals, reals, infinite cardinals and infinite ordinals. Similarly, there are many types of algorithms making a rigorous characterisation of algorithm either impractical or impossible (Gurevich, 2012a, p. 32).

In order to allow a critical evaluation of the algorithm execution account, we need to settle on a particular characterisation of an algorithm. For example, on Cummins' view, a physical system computes when its output and final state are a causal outcome of its input and initial state (1977, pp. 279–284, 1989, pp. 91–92, 1996, p. 610). By his lights, the execution of an algorithm reduces to a *disciplined step satisfaction*, which can be given a systematic semantic interpretation (Cummins, 1989, pp. 89–92). An algorithm then reduces to a systematically interpret*able* step satisfaction process, in which causal relations obtain among its steps as well as between its initial state and its final state.

---

[7] Clearly, it may be argued that recursion *is* a primitive operation, which is used and necessary for some very fundamental problems in computability theory (e.g., algorithms for binary search trees are defined recursively). Consider, for example, Cobham's theorem, according to which a function is primitive recursive *iff* it is computed by a program with primitive recursive runtime (Machtey & Young, 1978, pp. 53–54).

[8] Nevertheless, this is only a weak analogy, since, arguably, each of these types of numbers *has* a rigorous definition. The relationships among these types of numbers are well established.

On Cummins' view, computable functions that are satisfied by a process P executed in a system S specify causal relations between steps in P as well as between the final state of S and its initial state (1989, p. 92, 1996, p. 610). If P satisfies the functions f and g, and g takes f's output as input, then the satisfaction of step f will yield the satisfaction of step g. S is computing an addition function, for example, when the underlying steps it satisfies *could be* systematically interpreted as addition. Cummins takes an algorithm to be "perfectly general" and "[free] from resource constraints and from physical wear and tear" (2010, p. 39). But he denies that an algorithm is simply an input-to-output relation (or, say, an addends-to-sums relation in the case of an addition algorithm).

However, Cummins' analysis of computation is, arguably, too narrow and falls prey to Searle's trivialisation of digital computation (Copeland, 1996, p. 353). For instance, in the case of a virtual TM, which is *simulated* on a conventional digital computer, no *causal* relations obtain among the contents of the TM's virtual tape and its actions. Nonetheless, the virtual TM computes. Further, it is possible to show that on some formal description of Searle's wall, the label-bearing states *satisfy* the appropriate functions (in a manner that is systematically interpretable) and so, on Cummins' view, the wall computes!

In the ensuing discussion that follows, we shall examine Copeland's algorithm execution account. By his lights, an algorithm is any series of instructions that determines the behaviour of a computing system. An algorithm Al is a finite set of instructions such that, for some computing system CS, each instruction of Al calls for one or more primitive operations of CS to be performed, either unconditionally or if specific conditions, recognisable by CS, are met. Accordingly, the computation of CS is characterised as acting in accordance with an algorithm (Copeland, 1997, p. 696).

An algorithm, by Copeland's lights, may be either *classical* or *nonclassical*; our focus is on classical algorithms. An algorithm is *classical* if the function, whose arguments are inputs into the algorithm and whose values are the corresponding outputs, is Turing-computable. Hence, according to Copeland, an algorithm may be classical despite calling for the execution of primitive operations that no standard TM can perform, such as quantum operations. An algorithm is *nonclassical* if it yields the computation of some uncomputable function by the computing system (Copeland & Sylvan, 1999, pp. 54–55). Hypercomputation lies beyond "the reach of the universal Turing machine" (Copeland, 2002, p. 462) and beyond the scope of this book. Therefore, our present discussion shall be limited to computation as the execution of *classical* algorithms.

7.2.2　The Key Requirements Implied by the Algorithm Execution Account

The algorithm execution account identifies three key requirements for a physical system to perform digital computation.

- 1ˢᵗ Requirement: there exists a labelling scheme and a formal description, *SPEC*, of the architecture of the system and an algorithm specific to that architecture.
- 2ⁿᵈ Requirement: the system paired with the labelling scheme being an honest model of *SPEC*.
- 3ʳᵈ Requirement: having the capacity of being prepared in the requisite configurations that specify the problem to be solved and producing the requisite configurations that specify the solution.

The first requirement is the existence a labelling scheme L of the system S and a formal description *SPEC* of the architecture of S and an algorithm specific to that architecture.[9] This requirement clearly needs unpacking. Firstly, the labelling scheme designates certain parts of S as label bearers and provides the method of specifying the label borne by each label-bearing part at any given time (Copeland, 1996, pp. 337–338).

Secondly, the architecture is either a concrete or a conceptual structure of the system concerned, such as the computer organisation of a 3.4GHz quad-core Intel Core i7 iMac, in the former case, or the functional structure of a TM, in the latter case. Lastly, *SPEC* is a formal description of a particular functional architecture of S and an algorithm specific to this architecture. It can be defined either axiomatically or not (Copeland, 1996, p. 338).

Importantly, an adequate labelling scheme should not introduce any unintended temporal specificities into the theory. It should obtain for *any time step* and remain applicable throughout the operation of the computing system. A labelling scheme that only applies to a *particular* time interval and does not specify what labelling holds prior to or subsequent to that time interval is *incomplete* (Copeland, 1996, pp. 348–349). Consider, for example, a labelling scheme (1, 0) or, say, (high, low) that describes a pair of flip-flops. Suppose that the voltage across the first flip-flop is 10v and across the second one is 5v. This pair of flip-flops may be *labelled* (1, 0) and accordingly be described as representing the number two in a binary notation in a manner that is not limited only to any particular time interval.

As regards the algorithm concerned, on this account, it depends on the primitive operations that are supported by the particular architecture that is described by *SPEC*. An algorithm α being specific to an architecture does not merely entail that any computing system with such architecture can execute α, but also that each instruction call of α induces the performance of a sequence of one or more primitive operations available in the architecture. Accordingly, an algorithm that contains an explicit multiplication instruction *cannot be specific* to an architecture that has addition but *not* multiplication available as a primitive operation (Copeland, 1996, p. 337).

The second requirement, which complements the first one, is that *the pair (S, L) be an honest model of SPEC*. The pair (S, L) has to be a model that does not use

---

[9] Note that Copeland points out that the existence of such L and SPEC may *be* true irrespective of whether it is *known* to be true (1996, p. 338).

nonstandard interpretations of expressions in *SPEC*. This, according to Copeland, implies two necessary conditions. The first condition is that the interpretation associated with that model support counterfactuals about computation. That is, if *S* supports, say, an addition operation, then it has to be part of *SPEC* even if *S* never enters the state leading to the execution of the addition operation. The second condition is that the labelling scheme not be *ex post facto* (Copeland, 1996, pp. 350–351).

Let us examine these two conditions in turn. According to the first condition, the interpretation associated with the model of *SPEC* has to support assertions about the counterfactual behaviour of *S*. For example, the intended interpretation of a statement of the form "if *c* obtains, then do *y*" is that when *c* obtains it brings about the action *y*. This is tantamount to saying that *c* causes *y* or that the intended interpretation supports the counterfactual "if *c* had occurred, then *y* would have occurred". However, the truth-condition of such conditional statements cannot be formulated in terms of physical causation. For these conditional statements have to apply not only to physical hardware of real-world computing systems, but also to abstract machines, such as TMs. If we say that each action of a TM is the result of its configuration, the phrase "is the result of" cannot be substituted with the phrase "is caused by", since the TM is purely an abstract entity in which causation, strictly, does not apply (Copeland, 1996, p. 341).

The second condition is explained by way of an example that uses Searle's wall. On a nonstandard interpretation of computation, Searle's wall computes. However, this is the case only because the labelling scheme that describes the wall's computation is constructed *ex post facto*. As a result of the labelling scheme being *ex post facto*, the computation of the wall is limited to a *specific time interval* $[t_i, t_j]$ (for $i, j \in \mathbb{N}^+$ and $i < j$). That is, unintended temporal specificities are introduced into the model. On a Searlean interpretation of, say, the universal quantifier in "for any t_i the axioms should hold", it is restricted to some $[t_i, t_j]$ interval. But this is clearly not the *standard* meaning of the universal quantifier.

Let us consider an arbitrary program that the wall supposedly executes, say, the Wordstar program. According to Searle's Triviality thesis (as discussed in Chapter 4), there exists some method for correlating binary numerals with regions of the wall. For instance, if the wall has a high polymer content, then when the number of polymer chains that end in a given space is odd it is assigned the numeral '0', otherwise, it is assigned the numeral '1' (Copeland, 1996, p. 343). The description of the wall may then be arbitrarily correlated to the labelling scheme that describes a particular execution cycle of Wordstar. But that arbitrary description of the wall is "at most a passive 'scoreboard' and is no more an active participant in the computation than the scoreboard is an active player in a game of billiards" (Copeland, 1996, p. 348). Such a correlation does not make any *essential use* of features distinctive to the wall and so this correlation may be reapplied for *any* architecture-algorithm specification and for *any* physical system with a sufficiently large number of discernible parts (Copeland, 1996, p. 343).

The third requirement is the system having the capacity of being prepared in configurations that specify the problem to be solved and producing configurations that specify the solution. Subcomponents of S (which may be treated as black boxes) have to make available some minimal number of primitive operations. S needs to have sufficient resources to enable these primitive operations to be sequenced, not necessarily in a linear mode, in some predetermined way. These resources should include a provision for *feedback*, whereby the results of previous applications of primitive operations may be the input for further applications of those same operations. For example, the output of a black box may be the input to the very same black box, either immediately or after passing through some intervening sequence of operations. In other words, that black box may be *stateful*. S takes a finite time to execute any of its primitive operations, and if S is an abstract system, such as the TM, there is no lower limit on the time taken.[10] Also, while an algorithm consists of finitely many instructions, there is no upper limit on the number of such instructions (Copeland & Sylvan, 1999, p. 49).

7.2.3 The Algorithm Execution Account Evaluated

Copeland maintains that Turing's analysis provides the necessary, but not sufficient, conditions for a system to count as computational. To uphold the sufficiency of Turing's analysis, Copeland argues that a distinction should be drawn between standard and nonstandard interpretations of computation. On a nonstandard interpretation of a theory, the intended meanings of the terms of the theory are not respected. For instance, on a nonstandard interpretation, such as Searle's, of the formal specification of a computing system, it might turn out that walls and buckets of water are indeed computational.

Let us explain what a nonstandard interpretation means here by using Copeland's example concerning statements about the geography of European capitals. Suppose one assigns the numeral 1 as the referent of 'London' and the numeral 15 as the referent of 'Copenhagen'. Sentences of the form "X is north of Y" may be assigned truth conditions of the form "the referent of 'X' > the referent of 'Y'". On this interpretation, the sentence "Copenhagen is north of London" is true, but is no longer about Copenhagen, London or any other European capital. Similarly, the sentence "Copenhagen is the most northerly European capital" is only true if numbers greater than 15 are excluded from the domain of this model. Yet, this restriction introduces an unwanted specificity to the model (Copeland, 1996, pp. 346–347). This does not mean though that if these sentences are true under a *nonstandard* interpretation, then they are also true under the *standard* interpretation.

---

[10] On the other hand, on the Gandy-Sieg account, in a physical universe where there is an upper limit on the velocity of signal propagation, there is by implication a lower limit on the time taken to perform a physical operation.

Copeland's algorithm execution account includes paradigmatic digital computing systems as computational and excludes paradigmatic non-computing systems. It, therefore, meets the dichotomy criterion. The three requirements above block paradigmatic non-computing systems from being regarded as genuinely computational, as long as the labelling scheme is chosen under a "standard" interpretation in Copeland's terminology.[11] On a standard interpretation, the solar system, for instance, does not compute solutions to its own equations of motion. Whilst the planets of the solar system may follow the relevant law of motion, in a very loose sense of 'follow', they *do not* execute an algorithm. Kepler's law of motion, for example, is not a list of instructions each of which is performed by calling some primitive operations of a given architecture of the solar system.

If the solar system computes, then there exists some *SPEC* that specifies the algorithm it executes and the supporting architecture. Anyone claiming that the solar system computes some function *f*, has to describe the algorithm executed for computing *f* as well as the solar system's computational architecture (Copeland, 1996, pp. 338–339). Similarly, as observed above, Searle's wall only computes because the labelling scheme of the "computing system" is constructed *ex post facto*. Searle's physical state correlation to labels may be reapplied for *any* architecture-algorithm specification and for *any* physical system with a sufficiently large number of discernible parts. As remarked in Chapter 4, these parts could be either macroscopic or microscopic.

At the same time, conventional digital computers, iPads and TMs are classified as genuinely computational, on Copeland's account. Any conventional digital computer, *C*, is clearly classified as genuinely computational, since it meets the three requirements above. There exists a labelling scheme *L* and a *SPEC* of an architecture, say, von Neumann architecture, and algorithms that are specific to that architecture, say, the Mac OS-X operating system and other programs compiled into assembly language, such that the pair (*C*, *L*) is an honest model of *SPEC*.

Let us consider Copeland's simple case of such a *C* whose CPU consists of three 1-byte registers: an instruction register *I*, a data buffer *O*, and an accumulator *A*. The behaviour of *C* can be described by some set of axioms specifying the primitive operations supported by *C*'s architecture (Copeland, 1996, pp. 340–341). In this specification, 'Î', for example, reads 'the contents of register *I*' and "Ô ⇒ x" reads "the contents of register *O* becomes x". The set of axioms may include the following six operations as well as others for conditional branching, incrementing the instruction register and so on.

1. *Ax*1 (Copy) if Î = 00000001 DO (Â ⇒ Ô) {Copy the contents of register *O* to register *A*}
2. *Ax*2 (Simple Add) if Î = 00000010 DO (Â ⇒ Â + Ô) {Add the contents of register *O* to the contents of register *A*}

---

[11] Scheutz rejects this conclusion (1998) as discussed below.

3. *Ax3* (Multiply) if $\hat{I}$ = 00000011 DO ($\hat{A} \Rightarrow \hat{A} \times \hat{O}$) {Multiply the contents of register O by the contents of register A}
4. *Ax4* (Indirect Add) if $\hat{I}$ = 00000100 DO ($\hat{A} \Rightarrow \hat{A} + \tilde{O}$) {Add the contents of the register, whose *address* is stored in register O to the contents of register A}
5. *Ax5* (Conjunction) if $\hat{I}$ = 00000101 DO ($\hat{A} \Rightarrow \hat{A}$ && $\hat{O}$) {Perform a logical conjunction of the contents of register O and the contents of register A}
6. *Ax6* (Disjunction) if $\hat{I}$ = 00000110 DO ($\hat{A} \Rightarrow \hat{A} \parallel \hat{O}$) {Perform a logical disjunction of the contents of register O and the contents of register A}

The intended standard interpretation of a statement of the form "if c DO y" is that the occurrence of c leads to the action y. Such a statement supports the counterfactual "if c had occurred, then y would have occurred". Even if in the course of C's execution of some algorithm, the instruction 00000100 never enters I, the axioms above still specify which action would have occurred, if it had.

Since Copeland does not restrict an architecture to only apply to physical hardware, his account also classifies TMs as computational. The TM's functional architecture consists of a read/write head, a tape, a state register and a table of instructions, which is the algorithm specific to that architecture. The primitive operations supported by the TM's architecture include reading a symbol on the tape, shifting the head one place to the right or left and writing a symbol. Because Copeland classifies TMs as computational, he does not formulate the truth conditions of "if c DO y" statements in terms of physical causation. In the case of a TM, each action is completely determined as the result of the scanned symbol and the TM's state.

Whether FSA qualify as computational, on this account, is debatable. An FSA is conveniently describable as a directed graph that specifies its possible state transitions. But the *architecture* that may be attributed to the FSA is underdetermined. Its "architecture" consists of a "system" and some input "device". But unlike a TM, the precise structure of that system is unspecified. The system may simply be in any number of predefined states in a manner akin to the instructions table of the TM. The underdetermination of the FSA's architecture allowed Putnam to argue that every physical open system implements every FSA. For the same reason, Scheutz claims that every physical system, which consists of different states and exhibits state transitions depending on some input, might count as a "standard model" on Copeland's account (1998, p. 48). Nevertheless, since Copeland's account allows for an abstract architecture as well, we may conclude that the algorithm execution account also classifies FSA as genuine digital computing systems. The FSA's architecture consists of a read head, a read-only tape, a state register and a table of instructions.

It is questionable whether discrete connectionist networks qualify as computational according to the algorithm execution account. Copeland argues that his account applies equally well to connectionist networks where the "registers" of the computing system are realised in a highly distributed way (1996, p. 341). In the case of a connectionist network, *SPEC* consists of a parallel architecture, which has a particular pattern of connectivity and certain weights on the connections, and an algorithm specific to that architecture consisting of step-by-step applications of a certain propagation rule and a certain activation rule (Copeland, 1996, p. 337). Copeland describes the computation of a discrete connectionist network as a non-sequential execution of a classical algorithm.

Nevertheless, it is hard to see how this account can correctly classify discrete connectionist networks as computational. For one thing, the algorithm executed is said to be *specific* to a given machine architecture (physical or not), but that makes it a *program*. An algorithm, unlike a program, does not impose any particular data types and is insensitive to the particulars of the underlying machine used to "execute" it. Arguably, discrete connectionist networks do not execute *programs*. Some discrete connectionist networks, such as feedforward networks that only have a single capacity, say, the linear summation of n input neurones, simply process data and do not require instructions for their normal operation. Connectionist networks that have more than one capacity do require instructions for exercising the correct capacity. But even then, such networks do not execute *programs* in the sense used in computer science.

Copeland's account meeting the conceptual criterion and the taxonomy criterion hinges on whether or not *algorithm* execution is tantamount to *program* execution. Let us first consider the conceptual criterion. The algorithm execution account can explain the core concepts such as 'program', 'data', 'architecture' and 'virtual machine'. The last one can be explained by this account as a process for simulating programs whilst abstracting away the physical particulars of the underlying hardware. The concept 'architecture' is clearly foundational for the present account and as we have seen above, it is applicable to both abstract machines, such as TMs, and physical machines, such as conventional digital computers.

Prima facie, the algorithm being *specific* to a particular architecture, on this account, seems to be problematic for some digital computers. Conventional digital computers need not have, say, the multiplication operation supported by their architecture as a *primitive* operation (cf. axiom A*x3* above). In that case, an algorithm that performs multiplication cannot be *specific* to that architecture. This difficulty is addressed by stating that "[a] program calling for multiplications can run on such an architecture only because the compiler [...] replaces each multiplication instruction in the program by a series of addition instructions" (Copeland, 1996, p. 337).

Nevertheless, the response to this difficulty provides further evidence that, indeed, on this account computation is taken to be the execution of *programs*. If that is true, then the algorithm execution account fails to provide a clear distinction between an algorithm and a program. Of course, this is not surprising given the nontrivial difficulty of rigorously defining what an algorithm is as observed above. For this reason, 'algorithm' is, typically, only defined *informally* in computer science textbooks. For example, "[an algorithm is] a well defined sequence of steps that always finishes and produces an answer" (Hopcroft et al., 2001, p. 373). The implication for the algorithm execution account is that the conceptual criterion is only partially met, because 'algorithm' and 'program' are used interchangeably.

The algorithm execution account cannot fully meet the taxonomy criterion for a similar reason. The key differentiator, on this account, is the distinction between *acting in accordance with* an algorithm (or more precisely, a program) and *executing* an algorithm (or more precisely, a program). Special-purpose TMs and computers designed to solve a particular problem may be described as cases of the former type. UTMs and conventional digital computers may be described as cases of the latter type. So the algorithm execution account allows the classification of some computing systems as soft programmable or stored-program and the classification of less computationally powerful systems as program controlled. Yet, the explanatory power of this account is limited when it comes to taxonomising some systems that only perform trivial computations, such as basic logic gates or n-bit adders. For, clearly, they neither execute nor act in accordance with programs.

The algorithm execution account meets the miscomputation criterion. Although Copeland does not explicitly address the phenomenon of errors in computational processes, on his account, a miscomputation is explicable in terms of a certain condition in *SPEC* that obtains under the *standard* interpretation without yielding the corresponding action. For example, a miscomputation occurs when a computer enters a state in which its instruction register contains the Copy instruction 00000001 (cf. *Ax*1 above), but the intended copying of contents between the two registers does not take place. It is *not* that the relevant action does not occur because the computer happens to be in some specific time interval $[t_i, t_j]$. Under their intended interpretation, the axioms above entail conditional statements that are true *at any given time* during the *normal* functioning of the computing system (Copeland, 1996, p. 349). The system miscomputes, if it fails to perform some intended action *y* in response to some condition *c*, under the standard interpretation of *SPEC*.

Thus construed, Scheutz's challenge of the counterfactual-support condition can be met. Scheutz argues that counterfactual support is not actually required as a criterion of a system performing computation, for "even real systems under 'different environmental conditions' will not support counterfactuals" (1998, p. 48). He asserts that even a desktop computer would stop working correctly, if the computer were exposed to a strong magnetic field. The counterfactual support, which is implied by the second requirement discussed above, only obtains "during the normal functioning of the [computing] device" (Copeland, 1996, p. 349). This is also supported by the following observation.

"There is an inverse relationship between the precision demands inherent in an architecture and the reliability of the corresponding piece of hardware. For example, the slightest leakages of charge from the accumulator will cause the machine [...] to become grossly unreliable at its task of computing [... its] function [...] (In a strict sense *it will no longer compute this function* at all but some other function)" (Copeland & Sylvan, 1999, p. 62, italics added).

Arguably, the implementation criterion poses the greatest challenge to Copeland's algorithm execution account. Scheutz argues that this account still does not solve the core problem of how physical states should be grouped properly to correspond to abstract computational states implemented by the computing system (personal communication). To support this claim, Scheutz provides a Putnam-style proof showing that for every open system S and for every FSA describable by $SPEC$ there exists a labelling scheme L such that the pair (S, L) is an honest model of $SPEC$ (1998, p. 48).

This proof requires unpacking, so we shall just point out two objections to it. First, as Scheutz points out in a footnote, the FSA constructed for this proof can also compute infinite strings in a finite amount of time (1998, p. 48, fn. 12). That is, this FSA can perform hypercomputation. This seems physically and logically implausible. Second, Scheutz's proof makes use of Putnam's principle of non-cyclical behaviour, but it is not clear that it is justified. As already observed in Chapter 4, this principle is problematic in physical systems that are subject to external influences or noise.

The correspondence of physical states to computational states, according to Copeland's account, is accomplished through the application of the labelling scheme. The labelling scheme consists of two parts: firstly, the designation of certain parts of the computing system as label-bearers, and secondly, the method of specifying the label borne by each label-bearing part at any given moment in time. So as not to fall prey to a loose notion of states correspondence, in a manner akin to Putnam's Triviality theorem, Copeland imposes two further restrictions on the latter part. The first restriction is that the interpretation associated with $SPEC$ supports counterfactuals about computation. The second restriction is that the labelling scheme is not *ex post facto*. Even if we conceded that the algorithm execution account does not meet the implementation criterion, there would remain another avenue to correct this account. This would be the case, if digital computing systems were viewed as implementations of the functions they realise, as proposed by Scheutz (1999, p. 190). The spatiotemporal discreteness of a digital system still lends itself naturally to algorithmic descriptions.

Lastly, the algorithm execution account trivially meets the program execution criterion. It simply *identifies* digital computation with the execution of programs. Of course, as observed above, this comes at the price of excluding logic gates and discrete connectionist networks as non-computational.

In sum, the algorithm execution account is inadequate, if it identifies algorithms with programs. It would seem that it does. Further, it may be argued that this account subscribes to a semantic view of digital computation, because of its dependence on a standard *interpretation* of $SPEC$ (Scheutz, personal communication). However, the standard interpretation advocated by Copeland is simply a matter of causally individuating the identity of the computing system relative to the task it performs, rather than picking some arbitrary description that introduces unwanted spatiotemporal specificities. The existence of a labelling scheme does not imply observer relativity in the sense attributed to computation by Putnam and Searle. A clear strength of the algorithm execution account is that it uses *only* intrinsic and mathematical representations as revealed by the above example of an axiomatic specification.

7.3 The Mechanistic Account

7.3.1 Introduction

According to the mechanistic account, a digital computing system is a mechanism that can be ascribed the function of generating output strings from input strings in accordance with a general rule, or a mapping, that applies to all strings and depends on the input strings and, possibly internal states of the system, for its application (Piccinini, 2007, p. 516). This account is underpinned by the following three principles.

1. The vehicles processed in the course of computation are medium independent;
2. The function of the system is to process those vehicles; and
3. The system operates according to some rule(s).

Let us now elaborate on these three principles. According to the first principle, the vehicles processed in the course of computation could be implemented in a variety of ways, such as mechanically, electronically, magnetically or optically. Any given computation can be implemented in multiple physical media, provided that the media have sufficient degrees of freedom that can be appropriately accessed and manipulated (Piccinini & Scarantino, 2011, p. 8). According to the second principle, the teleological function of a computing system is to compute by processing medium-independent vehicles irrespective of their particular physical implementation (Piccinini, 2007, p. 507). According to the third principle, the operation of a computing system is performed in accordance with rules. These rules need not necessarily be programs as in the case of special-purpose TMs, FSA and connectionist networks (Piccinini, 2007, p. 518). A rule is defined as a mapping from input, plus possibly internal states, to output (Piccinini & Scarantino, 2011, p. 8).

Crucial to the mechanistic account is the notion of a *digit*. Computability only applies directly to abstract systems, such as TMs, Post machines and FSA, rather than physical systems. As observed in Chapter 3, computability is typically defined over strings of letters, often called symbols, from a finite alphabet. However, not every process that is defined over strings of letters counts as digital computation. A genuine random generation of a string of letters, for example, does not count as computation. On the mechanistic account, a *digit* is characterised as a concrete counterpart to the formal notion of a *letter* in computability theory (Piccinini, 2007, p. 510).

A digit is a discrete stable state of a component that is processed by the computing system. Piccinini originally defined it as "[*either*] a component *or* [a] state of a component of the mechanism that processes it" (2007, p. 510, italics added). As a component, "[i]t may enter the mechanism, be processed or transformed by the mechanism, and exit the mechanism" (ibid). Yet, the definition of a digit has later narrowed down to "a macroscopic state (of a component of the system) whose type can be reliably and unambiguously distinguished by the system from other macroscopic types" (Piccinini & Scarantino, 2011, p. 7). That is, a digit is a state but not a component. As also remarked in Chapter 3, according to Piccinini, digits need not represent anything including numbers. *Numeral digits* represent numbers, but other "mechanistic digits", such as "|" or "\", do not represent anything in particular. In a similar manner to TMs that only use finitely many symbols, there is an upper bound on the number of digit types a physical computing system can process (Piccinini & Scarantino, 2011, p. 8). Examples of digits are memory cells in conventional digital computers and holes, or lack thereof, on cards in punch-card computers.

Moreover, digits can be concatenated to form sequences or strings. Any relation between digits that has the abstractly defined properties of concatenation may constitute a concrete counterpart to abstract concatenation of letters in computability theory. Examples of ordering relations are spatial contiguity among digits, temporal succession among digits and a combination of both. Consider a conventional digital computer. Its components have to be arranged so that it is clear what the input digits are and what the output digits are. For the input digits and output digits to constitute strings, the components have to be arranged so as to respect the desired relations among the digits composing the strings. A four-bit adder circuit, which generates the binary sum of two four-bit numbers, has to be *functionally organised* so that the four digits in the input strings are manipulated in the correct order (Piccinini, 2007, p. 512)

7.3.2   The Key Requirements Implied by the Mechanistic Account

The mechanistic account identifies four key requirements for a physical system to perform digital computation.

- 1st Requirement: processing tokens of the same digit type in the same way and tokens of different digit types in different ways.
- 2nd Requirement: processing all digits belonging to a string during the same functionally relevant time interval and in a way that respects the ordering of the digits within that string.
- 3rd Requirement: any system component that processes digits stabilises only on states that count as digits.
- 4th Requirement: the system's components have to be functionally organised and synchronised so that external inputs, together with any digits stored in memory, have to be processed by the relevant components in accordance with a set of rules.

The first requirement centres on the manner in which digit tokens are processed by the system. Digits are permutable in the sense that typically any token of any digit type may be replaced by a token of any other digit type (Piccinini, 2007, p. 510). Strings of digits can be either data or rules, so they only differ in the functional role they play during processing by the computing system (Piccinini & Scarantino, 2011, pp. 7–8). Under normal conditions, digits of the same type in a computing system affect primitive components of the system in sufficiently similar ways, thereby, their dissimilarities make no difference to the output produced. On the other hand, the differences between digit types have to suffice for the processing components to differentiate between these types, so as to yield the correct outputs.

Consider a conventional two-input, one-output XOR gate. The two inputs to the gate that are sufficiently close to a specific voltage, labelled type '1', yield an output of a voltage sufficiently close to a different specific value, labelled type '0'. However, that does not imply that for *any* two input types the primitive computing component *always* yields outputs of different types. Tokens of two different input types can yield the same computational output, as in the case of a NOR gate. The three input types (1, 1), (0, 1) and (1, 0) – all give rise to same '0' output type. Nevertheless, it is essential that the NOR gate yields different responses to tokens of different types, thus responding to input types (0, 0) differently from the other possible input types.

The second requirement centres on the spatiotemporal ordering of digits processed by the system. When a computing system is sufficiently large and complex, there has to be some way to ensure synchronisation among all digits belonging to a particular string. Most complex computing systems, including logic circuits, are parallel in the sense that they perform more than one computational operation during any relevant time interval (Piccinini, 2008a, p. 53). This requirement is related to the fourth requirement of the Gandy-Sieg account examined above. The components of a computing system interact over time and given their physical characteristics, there is only a limited amount of time during which their interaction can produce the correct result, which is consistent with the spatiotemporal ordering of digits within strings.

The type of the digit ordering within strings that affects their processing varies between primitive and complex computing systems. In primitive computing components and simple circuits it is mostly the temporal ordering of digits that is crucial for producing the correct result. For example, if digits that were supposed to be summed together entered a four-bit adder circuit at times that are *too far apart*, they would not be added correctly (Piccinini, 2007, p. 513). In more complex systems, the processing of all digits belonging to a string has to proceed in a way that also respects the spatial ordering of the digits within the string. Each digit in the sequence has to be processed until the last digit in the string is reached. In some cases the spatial ordering of digits makes no difference to the computational process. For example, at the program level, for both the summation of numbers in an array and a calculation of the length of a sequence of symbols, the spatial ordering of digits does not affect the correctness of the output produced.

The third requirement centres on the system's components stabilising on states that count as digits. Components of digital systems can be in any one of several stable states. In a binary computing system, memory cells, for instance, can be in either of two thermodynamically macroscopic stable states, each of which constitutes a digit. Upon receiving some physical stimulus, such as a reboot of a desktop computer or the depressing of a key on a keyboard, the memory cell enters a state on which it stabilises. The memory cells stabilise on states corresponding to one of two digit types, which are typically labelled '0' and '1', that are processed by the computing system. If these memory cells lost their capacity to stabilise on one of these two digit types, they would cease to function properly and the binary computing system would cease to function normally (Piccinini, 2007, p. 511).

The fourth requirement centres on the processing of digits in accordance with a set of rules. During each time interval, the processing components transform external input, if such exists, and previous memory states in a manner that corresponds to the transition of each computational state to its successor. Any external input combined with the initial overall memory state constitute the initial string of a particular computation. Intermediate memory states constitute the relevant intermediate strings. Similarly, the output produced by the system and the final memory state combined constitute the final string of the computation. As long as the components of the system are *functionally organised* and *synchronised* so that their processing of digits respects the well-defined ordering of the digits concerned, the operation of the system can be described as a series of snapshots. The set of computational rules specifies the relationship that obtains between inputs and the respective outputs produced by modifying snapshots (Piccinini, 2007, pp. 509, 515). In systems, which Piccinini dubs "fully digital" computing systems (personal communication), these snapshots are modified in accordance with *programs*. Other computing systems, such as some discrete connectionist networks, take digital inputs and produce digital outputs in accordance with rules defined over these inputs without executing programs (Piccinini, 2008b, p. 320).

7.3.3 The Mechanistic Account Evaluated

Given that Piccinini explicitly evaluates the mechanistic account based on most of the adequacy criteria discussed in Chapter 2, the ensuing discussion follows his evaluation whilst emphasising any contentious points.

The mechanistic account meets the dichotomy criterion for two reasons. The first one is that:

> "[a]ll paradigmatic examples of computing mechanisms, such as digital
> computers, calculators, Turing machines, and finite state automata, have
> the function of generating certain output strings from certain input strings
> and internal states according to a general rule that applies to all strings

and depends on the inputs and internal states for its application. According to the mechanistic account, then, they all perform computations" (Piccinini, 2007, pp. 517–518).

The extensional class of digital computing systems denoted by the mechanistic account hinges on how 'mechanism' is construed. If a mechanism is taken literally as a spatiotemporal system, then FSA, CSA, TMs and similar abstract computational models are excluded from this class. However, Piccinini wants to classify FSA and TMs as genuine digital computing systems. Therefore, a mechanism, by his lights, is construed as either an abstract or a concrete entity depending on the context of enquiry.

Although they are not the paradigmatic computing systems, discrete connectionist networks are also classified by the mechanistic account as computational. For "they manipulate strings of digits in accordance with an appropriate rule" (Piccinini, 2008b, p. 316). The inputs to their input units and the outputs from their output units are digits, because they "are relevant only when [these units] are in one of a finite number of (equivalence classes of) states" (Piccinini, 2008b, p. 316). And since connectionist networks can be decomposed into units with functions and an organisation, they are classified as *mechanisms* (Piccinini, 2007, p. 518). Piccinini concludes "that connectionist systems that process continuous variables do something other than computing, at least in the sense of 'computing' employed in computer science and computability theory" (2008b, p. 319). As argued in the next chapter, connectionist networks of this type perform *analogue* computation.

The second reason that this account meets the dichotomy criterion is that it also excludes paradigmatic cases of non-computing systems by appealing to their mechanistic explanation or lack thereof. Consider the following cases of paradigmatic non-computing systems that are excluded by the mechanistic account.

"[P]*lanetary systems* and the weather—are not mechanisms in the present sense, because they are not collections of functional components organized to exhibit specific capacities. Also, most systems—including, again, planetary systems and *the weather*—do not receive inputs from an external environment, process them, and return outputs distinct from themselves" (Piccinini, 2007, p. 520, italics added).

"[T]here is no prominent scientific theory according to which *digestion* is a computational process. [... T]he current science of digestion does not identify finitely many types of input digits and an ordering between them, let alone processes that manipulate those inputs in accordance with a rule defined over the input types and their place within strings. [... W]hat matters to digestion are not the details of how molecules of different chemical types, or belonging to different families, form pieces of food" (Piccinini, 2007, p. 521, italics added).

"[N]ot all mechanisms that manipulate strings of symbols do so in accordance with a general rule that applies to all strings and depends on the input strings for its application. [...] A genuine *random* [...] *numeral generator* produces a string of digits, but it does not do so by computing, because there is no rule for specifying which digit it will produce at which time. Thus, a genuinely random [... numeral] generator does not count as a computing mechanism" (Piccinini, 2007, p. 522, italics added).

The mechanistic account of computation meets the conceptual criterion as well. Compilers and interpreters, on this account, play a role within the functional hierarchy of the computer in the process of executing programs. All the relevant elements of the computing process, including operating systems, compilers, interpreters and virtual machines, can be functionally characterised as strings of digits operated upon by the processor. These mechanisms are identified as "nothing but programs executed by the computer's processor(s)" (Piccinini, 2008a, p. 46). Still, they give rise to a hierarchy of functional organisations within the *stored-program* computer by generating virtual memory, complex notations and complex operations.

Whilst *data* are not explicitly defined in the mechanistic account, they are treated as *strings*. Piccinini distinguishes three classes of strings on the basis of the teleological function they fulfil.

1. *Data,* whose function is to be manipulated during a computation;
2. *Results,* whose function is to be the final string of a computation;
3. *Instructions,* whose function is to cause appropriate computing mechanisms to perform specific operations on the data (Piccinini, 2008a, p. 34).

This classification is problematic. Firstly, data are distinguished from instructions and results. Secondly, data are classified *passively* and treated as raw data operated upon by the computing system. The same applies to the classification of the results. It is only the class of instructions that has a genuinely *active* function within the computing system. Elsewhere, Piccinini asserts that the very *same* string of digits may play the role of an instruction in one run of the program and data in another (Scarantino & Piccinini, 2010, p. 326). He maintains that many computations "depend not only on an input string of *data,* but also on the *internal state* of the [...] mechanism" (Piccinini, 2007, p. 509). So, *data* seem to be distinguished from *states* but at the same time, on the mechanistic account, they qualify are strings. This shortcoming reveals an important advantage of the IIP account that provides a richer notion of data, thereby allowing an explanation of significantly more computational properties (Fresco & Wolf, forthcoming).

The mechanistic account meets the program execution criterion too. It rejects the equivalence of digital computation and program execution. Instead, on this account, program execution is considered a special case of computation. Program execution is the process by which a certain state of the mechanism affects the processing component of the mechanism, so as to perform sequences of operations. But digital computation is not reducible to program execution. Some automatic looms, for example, operate, according to Piccinini, by executing programs, but they do not digitally compute. Program-executing computing systems, but not automatic looms, are defined over, and responsive to, the finitely many input types they process and the order of processing inputs (Piccinini, 2007, p. 517). Program execution is a property of only *some* computing systems – those that are soft programmable.

Unlike other accounts of concrete digital computation, the mechanistic account explicitly addresses the phenomenon of computational errors. It explains "what it means for a computing mechanism to make a mistake" (ibid: p. 523). But it should be clear from the outset that mistakes are only one type of miscomputation. For mistakes are related to the description and design of the problem to be solved by computational means, whereas miscomputation broadly construed can also be the result of hardware malfunction due to physical noise, such as the meltdown of some component (as a result of overheating) or an extreme exposure of the computing system to some electromagnetic field. As observed in Chapter 2, the definition of miscomputation, on the mechanistic account, is too narrow. With that caveat in mind, let us examine why the mechanistic account meets the miscomputation criterion.

A miscomputation, according to the mechanistic account, is an event in which a working computing system fails to fulfil its function resulting in a computational error. Three main types of miscomputation are identified. The first, most important, type of miscomputation, for which a physical computing system can be "blamed", is a hardware malfunction. It is explained on this account as the failure of some component to perform its function within the overall computing mechanism. Such a malfunction could be the failure of either a computing component, say, a logic gate, or a non-computing component, such as a clock or a battery. Another type of miscomputation, a program error, may be attributed to either a mistake in the design specification of the computer program or an instruction that cannot be executed, say, due to a memory overrun or incorrect memory addressing. A third type of miscomputation may be the result of round-off errors accumulating in the finite precision arithmetic that computer processors employ (Piccinini, 2007, pp. 523–524).

The mechanistic account of computation also meets the taxonomy criterion. It classifies "genuine [digital] computers" as being "programmable, stored-program, and computationally universal" (Piccinini, 2007, p. 524). It construes programmability, stored-programmability and universality as functional properties that can be explained mechanistically. Different computing systems may be distinguished based on these functional properties. Conventional digital computers have all these properties. Special-purpose computers and physical instances of special-purpose TMs, on this account, do not have any of these properties (Piccinini, 2008a, p. 55). Such systems can be classified, as suggested in Chapter 3, as being program-controlled, that is, systems that act in accordance with "hardwired" programs.

The mechanistic account also allows the taxonomy of other computing systems, such as nonprogrammable calculators, logic gates and logic circuits. Although these systems are not programmable, nor stored-program nor computationally universal, they compute nonetheless. Logic gates perform trivial operations on bits. Nonprogrammable calculators can compute a finite but not an insignificant number of functions for certain inputs of bounded size. On the mechanistic account, logic gates cannot be further decomposed into simpler computational subcomponents, so they are classified as "primitive computing components". Because nonprogrammable calculators manipulate strings of digits according to a general rule, they are also classified as computing systems, albeit of limited computational power.

Nevertheless, the taxonomy offered by the mechanistic account seems to suffer from two shortcomings. Firstly, Piccinini argues that UTMs should not be treated as stored-program computing systems (2008a, pp. 55–56). This leads to the curious result that whilst a conventional digital computer has all three capacities above, computationally – it should be "ranked" below a UTM, since the former may run out of memory or time. Secondly, on the mechanistic account, FSA are "so much less [computationally] powerful than ordinary digital computers" (Piccinini, 2007, p. 506). However, since ordinary computers are basically large complicated FSA, the mechanistic account may not be effective at separating complexity due to size, as is the case with computers and FSA, from complexity due to a particularly rich structure, as is the case with combinational circuits and flip- flops (Fresco & Wolf, forthcoming).

Whether or not the mechanistic account is ultimately judged adequate hinges on meeting the implementation criterion. On this account, digits and strings of digits are equivalence classes of physical states. Given their special functional characteristics, digits can be labelled by letters and strings of digits by strings of letters. Consequently, the same formal operations and rules that define abstract computations over strings of letters can be used to characterise concrete computations over strings of digits. The input received from the environment and the initial overall memory state of the computing system implement the initial string of an abstract computation (Piccinini, 2007, pp. 514–515).

But how does such an implementation relation obtain? Piccinini offers the following analysis.

"Abstract computations can be reduced to elementary operations over individual pairs of letters. Letters may be implemented as digits by the state of memory cells. Elementary operations on letters may be implemented as operations on digits performed by logic gates. Logic gates and memory cells can be wired together so as to correspond to the composition of elementary computational operations into more complex operations. Provided that (i) the components are wired so that there is a well-defined ordering of the digits being manipulated and (ii) the components are synchronized and functionally organized so that their processing respects the ordering of the digits, the behavior of the resulting mechanism can be accurately described as a sequence of snapshots. Hence, under normal conditions, such a mechanism processes its inputs and internal states in accordance with a program; the relation between the mechanism's inputs and its outputs is captured by a computational rule" (Piccinini, 2007, p. 515).

Whilst this analysis does not shed light on how physical states are grouped together to correspond to abstract computational states, the correspondence relation is neither arbitrary nor *ex post facto* like Searle's proposed labelling of physical states. Even if we adopted the strict implementation criterion suggested by Scheutz, his proposed alternative of implementation as the "realisation of functions" is certainly compatible with the mechanistic account. For it is the case that, on the mechanistic account, every computational subcomponent of a computing system is attributed some function it has to fulfil so that the overall mechanism be classified as genuinely computational.

In sum, the mechanistic account of computation is evaluated positively overall based on our recommended adequacy criteria. Additionally, it does not appeal to extrinsic representations for explaining concrete computation. The main notions of digits and strings of digits, which are employed by this account, are functional and non-semantic. They are explicated in terms of how the type of each digit and its spatiotemporal ordering within a string affect their processing by the computing system. The only type of representation that is required by this account for explaining the "computer's understanding" of the instructions it processes is intrinsic. This is not problematic.

7.4 Concluding Remarks

We have seen above that none of the three accounts analysed in this chapter posits any extrinsic representational properties. This makes them preferable to semantic accounts of digital computation. The representational character of concrete digital computation is further examined in the first part of Chapter 8. Only the mechanistic account is found adequate, on the present analysis. Being an adequate account of computation proper, it enables cognitive scientists to demonstrate which aspects of cognition, if any, are computational based on the features it identifies. In Chapter 8, we examine the implications of this account for digital computationalism in the context of cognitive science.

References

Copeland, B.J.: What is computation? Synthese 108(3), 335–359 (1996), doi:10.1007/BF00413693

Copeland, B.J.: The Broad Conception of Computation. American Behavioral Scientist 40(6), 690–716 (1997), doi:10.1177/0002764297040006003

Copeland, B.J.: Hypercomputation. Minds and Machines 12(4), 461–502 (2002), doi:10.1023/A:1021105915386

Copeland, B.J., Shagrir, O.: Physical Computation: How General are Gandy's Principles for Mechanisms? Minds and Machines 17(2), 217–231 (2007), doi:10.1007/s11023-007-9058-2

Copeland, B.J., Sylvan, R.: Beyond the universal Turing machine. Australasian Journal of Philosophy 77(1), 46–66 (1999), doi:10.1080/00048409912348801

Cummins, R.: Programs in the Explanation of Behavior. Philosophy of Science 44(2), 269–287 (1977), doi:10.1086/288742

Cummins, R.: Meaning and mental representation. MIT Press, Cambridge (1989)

Cummins, R.: Systematicity. The Journal of Philosophy 93(12), 591–614 (1996), doi:10.2307/2941118

Cummins, R.: The world in the head. Oxford University Press, Oxford (2010)

Davies, E.B.: Building Infinite Machines. The British Journal for the Philosophy of Science 52(4), 671–682 (2001), doi:10.1093/bjps/52.4.671

Dowek, G.: Around the Physical Church-Turing Thesis: Cellular Automata, Formal Languages, and the Principles of Quantum Theory. In: Dediu, A.-H., Martín-Vide, C. (eds.) LATA 2012. LNCS, vol. 7183, pp. 21–37. Springer, Heidelberg (2012)

Fresco, N., Wolf, M.J.: The instructional information processing account of digital computation. Synthese (forthcoming), doi:10.1007/s11229-013-0338-5

Gandy, R.: Church's Thesis and Principles for Mechanisms. In: The Kleene Symposium, pp. 123–148. North-Holland (1980)

Gibbs, N.E., Tucker, A.B.: A model curriculum for a liberal arts degree in computer science. Communications of the ACM 29(3), 202–210 (1986), doi:10.1145/5666.5667

Gurevich, Y.: Algorithms in the world of bounded resources. In: Herken, R. (ed.) A Half-century Survey on the Universal Turing Machine, pp. 407–416. Oxford University Press, New York (1988)

Gurevich, Y.: What Is an Algorithm? In: Bieliková, M., Friedrich, G., Gottlob, G., Katzenbeisser, S., Turán, G. (eds.) SOFSEM 2012. LNCS, vol. 7147, pp. 31–42. Springer, Heidelberg (2012a)

Gurevich, Y.: Foundational Analyses of Computation. In: Cooper, S.B., Dawar, A., Löwe, B. (eds.) CiE 2012. LNCS, vol. 7318, pp. 264–275. Springer, Heidelberg (2012b)

Hopcroft, J.E., Motwani, R., Ullman, J.D.: Introduction to automata theory, languages, and computation. Addison-Wesley, Boston (2001)

Machtey, M., Young, P.: An introduction to the general theory of algorithms. North-Holland, New York (1978)

Moschovakis, Y.N.: On founding the theory of algorithms. In: Dales, H.G., Oliveri, G. (eds.) Truth in Mathematics, pp. 71–104. Clarendon Press, Oxford (1998)

Moschovakis, Y.N., Paschalis, V.: Elementary Algorithms and Their Implementations. In: Cooper, S.B., Löwe, B., Sorbi, A. (eds.) New Computational Paradigms, pp. 87–118. Springer, New York (2008)

Piccinini, G.: Computing mechanisms. Philosophy of Science 74(4), 501–526 (2007), doi:10.1086/522851

Piccinini, G.: Computers. Pacific Philosophical Quarterly 89(1), 32–73 (2008a), doi:10.1111/j.1468-0114.2008.00309.x

Piccinini, G.: Some neural networks compute, others don't. Neural Networks 21(2-3), 311–321 (2008b), doi:10.1016/j.neunet.2007.12.010

Piccinini, G., Scarantino, A.: Information processing, computation, and cognition. Journal of Biological Physics 37(1), 1–38 (2011)

Scarantino, A., Piccinini, G.: Information without truth. Metaphilosophy 41(3), 313–330 (2010), doi:10.1111/j.1467-9973.2010.01632.x

Scheutz, M.: Do Walls Compute After All? Challenging Copeland's Solution to Searle's Theorem Against Strong AI. In: Proceedings of the 9th Midwest AI and Cognitive Science Conference, pp. 43–49. AAAI Press (1998)

Scheutz, M.: When Physical Systems Realize Functions... Minds and Machines 9(2), 161–196 (1999), doi:10.1023/A:1008364332419

Shagrir, O.: Effective Computation by Humans and Machines. Minds and Machines 12(2), 221–240 (2002), doi:10.1023/A:1015694932257

Sieg, W.: On mind & Turing's machines. Natural Computing 6(2), 187–205 (2007), doi:10.1007/s11047-006-9021-9

Sieg, W.: Church Without Dogma: Axioms for Computability. In: Cooper, S.B., Löwe, B., Sorbi, A. (eds.) New Computational Paradigms, pp. 139–152. Springer, New York (2008)

Sieg, W., Byrnes, J.: An abstract model for parallel computations: Gandy's thesis. The Monist 82(1), 150–164 (1999)

Turing, A.M.: On Computable Numbers, with an Application to the Entscheidungsproblem. Proceedings of the London Mathematical Society s2-42(1), 230–265 (1936), doi:10.1112/plms/s2-42.1.230

Chapter 8
Computation Revisited in the Context of Cognitive Science

In this final chapter, we conclude the book with four theses. First, in Section 8.1, we argue that in the course of nontrivial computation, typically, only *implicit* intrinsic and mathematical representations are processed. Then, in Section 8.2, we argue that any computational explanation of cognition is unintelligible without a commitment to a single interpretation of 'digital computation' as defined by a given account thereof. In Section 8.3, we argue that a blanket dismissal of the key role computation plays in cognitive science is unwarranted. We also argue that the thesis that computationalism, connectionism and dynamicism are mutually exclusive is wrong. Finally, in Section 8.4, we conclude with some reflections on the computational nature of cognition and suggest some future research opportunities.

8.1 A Limited Representational Character of Digital Computation

We have seen in the previous chapters that some accounts of concrete digital computation posit extrinsic representations, whilst others do not. To reprise the distinction made in Chapter 1 between intrinsic and extrinsic representation, the former is "confined" to the physical boundaries of the computing system, whilst the latter is not. Mathematical representation, which is simply a representation of a function to be evaluated, does not compete with either intrinsic or extrinsic representations. The FSM and PSS accounts examined in Chapter 5 emphasise the extrinsic representational character of digital computation. But this is not surprising. For the proponents of these accounts have a more ambitious goal than "merely" characterising digital computation, namely, advancing a substantive empirical hypothesis that human cognition is essentially a computing system of the specified sort. According to Smith's participatory account, digital computing systems traffic in extrinsic representations.

Still, other accounts examined in Chapters 6 and 7 do not posit extrinsic representations. The IIP account does not appeal to extrinsic representations for explaining concrete digital computation, since it is not based on *factual* information, which, as construed in Section 6.2.3, has to be true, according to the

Veridicality thesis. The Gandy-Sieg account, the algorithm execution account and the mechanistic account also do not appeal to extrinsic representations for explaining concrete digital computation. Our objective has been to explain digital computing systems as being *computational* rather than as being *representational*. For that reason, these four accounts are preferable, at least prima facie, to accounts that *do* posit extrinsic representations. Whilst digital computation need not be individuated by extrinsic representations, it is individuated by intrinsic and mathematical representations.

In the remaining section, we discuss the following argument.

- (P1) The types of representation that are essentially processed in the course of computation are intrinsic and mathematical.
- (P2) For a representation to be explicitly encoded it has to be readily available.
- (P3) The representations processed in the course of *nontrivial* computation are, typically, not readily available.
- Therefore, (C) In the course of nontrivial computation, typically, only *implicit* intrinsic and mathematical representations are processed.

Why is the first premise (P1) plausible? The short answer is that concrete digital computation can, but need not, process any extrinsic representations. According to the classification of computational representations offered in Chapter 1, there remain two other types of representation to be considered: intrinsic and mathematical. Suppose that a deterministic conventional computing system S is programmed with particular axioms and rules for inferring "new" propositions from "old" ones. Following the transformation of some symbolic expressions, initially only the axioms, S produces as output a specific proposition p, which corresponds to some state of affairs in world W_X, that is, p has empirical content. Suppose that S is then placed in another possible world W_Y different in some regards from W_X, but with the same laws of physics. S should produce the same p when S is given the same input and its initial state remains unchanged. Yet, p in world W_Y may not be true anymore, since it fails to correspond to the same state of affairs that holds in world W_X. Should we say that when producing p, S computes in W_X but miscomputes in W_Y? That hardly seems plausible. The data processed by S may qualify as factual information in some cases (e.g., when p extrinsically represents some state of affairs in W_X), but their empirical content is irrelevant to their processing in the course of computation.

It helps to compare abstraction in mathematics and abstraction in computer science to better understand the role of intrinsic and mathematical representation in the course of computation. Both mathematical discourse and computer science discourse fundamentally include abstract objects, but they differ in the nature of these objects (Colburn, 1999, p. 10). "The propositions of mathematics are devoid of all factual content; they convey no information whatever on any empirical subject matter" (Hempel, 2000, p. 13). The nature of mathematical abstraction is to remove the semantics of particular terms. Pure deduction proceeds with terms, or axioms, that have no external semantics, though they *may* happen to describe physical reality (Colburn, 1999, p. 11). In our terminology, propositions in mathematics can, but need not, be extrinsic representations.

The nature of abstraction in computer science, on the other hand, is to give the particular terms more expressive power (Colburn, 1999, pp. 12–14). At the physical level, a computational process realised by a memory-based logic circuit is a sequence of changes in the states of the circuit, where each state is described by the presence or absence of, say, electrical charges in memory and processing elements. *Procedural* abstraction, enabled by high-level programming languages, and *data* abstraction introduce (abstract) objects such as instructions, procedures, variables and data types. These abstractions are placed on the physical states of memory elements conventionally *interpreted* as numbers. Nevertheless, abstract objects, at the program level, are intimately tied to particular machine implementations, at the physical level.

Whilst the data processed by the computing system might have external semantics, such external semantics is irrelevant to their computational processing. The computing system would compute the appropriate functions irrespective of any *external* semantics the data possibly have. The expressive power that is gained by way of abstraction in computing systems is limited to *formal* semantics, such as operations on numbers, strings, bit arrays and sets as well as distinguishing "pure" data from instructions.

Consider procedural abstraction that is used in high-level programming languages. Entities on all scales of a computer program get formal semantic values. Variables and constants of the program change their values through assignment. But assignment statements, procedures (or subroutines) and even the whole program are also assigned some semantic values. That programs have semantic values on all scales is a basic principle of denotational semantics. Procedures are invoked with particular values of their parameters – that is, data – and perform various actions on them by way of following instructions. A procedure, P, can be conceived of as representing a mathematical function, whose arguments are the semantic values of P's parameters, and whose value is the semantic value returned by P. The semantic value of a procedure maps argument types to a return type. A procedure, P_1, can also take another procedure, P_2, as a parameter and return a value (White, 2004, p. 239). This is a clear example of intrinsic representation whereby P_2 is represented by the parameter that is passed to P_1. The nesting of procedures is fundamental to recursion and is very common in functional programming as well as in object-oriented programming (in passing an "object" as a parameter – its procedures, typically called methods, are also passed).

Programmable digital computing systems are driven by programs that make regular use of intrinsic representation: one procedure is represented by another, some object is represented by another and one software library is represented by another. The motivation is that if two items "behave" in the same manner, in an informal sense, then they are interchangeable. Specific semantic properties of programming languages make this referential transparency possible (White, 2004, p. 241).

But even at the hardware level, a system computes by means of intrinsic and mathematical representations. For example, in a conventional digital computer, the CPU shifts data and instructions for execution between the computer's main memory and its registers in the course of executing primitive operations. Some instructions contain addresses of registers in the main memory and in turn these registers may contain representations of other instructions. The part of the instruction that is fed into the control unit of the CPU can be viewed as a mathematical representation, since it determines some mathematical operation to be performed, such as add the content of two registers and store the result in a third register. Of course, none of these examples requires an appeal to extrinsic representations for understanding the computation performed.

Consider lastly the example of a four-bit adder circuit. This circuit adds two four-bit addends and produces their sum, thereby representing the addition function on two addends. The circuit operates at the electrical level by manipulating voltage levels on its input lines and sending the "resulting" voltage pulse to its output line. At a *purely* physical level there is no need to appeal to *any* type of representation to explain its operation. However, a good explanation of the computation the circuit performs has to appeal to intrinsic and mathematical representations. The circuit takes representations of two addends and produces a representation of their sum. The adder circuit is designed so that, under the right conditions, the pulse sent to its output line has a conventional interpretation that corresponds to the sum of the two numbers represented by the pulses on its input lines. The appeal to such representations is required to determine whether the circuit *computes* or *miscomputes*.

Let us now examine the second premise (P2). What does it take for a representation to be explicitly encoded? Is a specific word in a book without an index considered explicitly encoded? If not, does it become explicit if it is indexed? Suppose that a set of axioms is represented in first order predicate logic, is the whole deductive closure of the set explicit? If so, is it also explicit when the deductively closed set is infinite? "That such questions arise is proof that we have unsettled intuitions about the meaning of explicit and implicit information" (Kirsh, 1991, p. 341). Our intuitions about explicitness and implicitness of representation are inconsistent. For instance, the numeral '5' in the set $S=\{3, 4, 1, 31, 11, 5, 9, 7, 2, 6\}$ seems to be explicitly represented in S. For the information that '5' is a member of S is "on the surface" of the set. From a structural perspective, a representation is explicit when it has a *definite location* and a *definite meaning*. From a process perspective, it is explicit when it is *ready to be used*. Whereas the numeral '5' appears to be in a usable form in S from a structural perspective, it is not immediately usable from a process perspective since it requires a computational search.

In the context of computational processes, a representation is explicitly encoded, if it is readily available (Kirsh, 1991). A computational representation is explicit, if it can be accessed, or activated, in constant time (Kirsh, 1991, p. 358). Constant time is the smallest complexity order known in computer science (also denoted as $O(1)$ time). It describes the execution time of an algorithm that is bounded by some value that is independent of the size of the input data. Consider, for example, the following simple procedure.

```
boolean isNumberEven(int iNumber){

    if (iNumber & 1)

            return false;

    else

            return true;

}
```

Whilst the exact execution time of *isNumberEven()* depends on the architecture of the CPU that computes the compiled code, it is upper bounded by a constant, irrespective of the size of *iNumber*. On the other hand, the execution time of a procedure that calculates the sum of all the numerical values stored in an array does depend on the size of the input, since the procedure has to traverse all the populated cells of the array to calculate the sum.

The "constant time criterion" places a precise operative constraint on the explicitness of representation, replacing other vague intuitions about the representation being *immediately usable* or *easily identifiable*. Of course, it is useful only if what can be done in a single computational step is known. In some digital computing systems, for instance, some data can be retrieved in just a few steps regardless of the size of the memory storage unit (Kirsh, 1991, p. 357). In such cases, there is no crucial difference between the time spent to *locate* data and the time spent to *decode* data. Both operations require algorithmic processing to perform either some form of pattern matching or data decryption (Kirsh, 1991, p. 344).

According to the third premise (P3), the representations processed in the course of *nontrivial* computation are, typically, not readily available. This premise requires some elucidation. Generally, the complexity of an algorithm is stated as a function relating the size of its input to the number of steps needed to compute the function. For example, the complexity of sorting algorithms, which compare the elements of an unsorted list, is proportional to the size of the input multiplied by its logarithm for element comparison for most inputs. The complexity of binary search algorithms for locating a particular value in a *sorted* list of values is proportional to a logarithm of the input's size.

In practice, very few nontrivial computations fall into the category of constant time complexity. For finite inputs, complexity as a measure is typically less useful, because if the answers to the problem are known, they can be stored in a look-up table, where only a minimal number of steps will be required to locate any particular answer (Kirsh, 1991, p. 357). A reduction of runtime complexity requires an increase of memory storage. If a giant look-up table for storing a finite set of answers to a given problem, the size of the memory storage needed grows in proportion to the input's size. However, this simply means that the algorithmic processing is essentially replaced by the look-up table search. Whether this look-up table qualifies as *computing* the answers to the original problem is debatable.

In sum, if we adopt the "constant time criterion" for the explicitness of representation, it follows from premises (P1)-(P3) that in the course of *nontrivial* computation, typically, only *implicit* intrinsic and mathematical representations are processed.

8.2 Avoiding Ambiguity about Computation

We now turn to draw the moral from our examination of extant accounts of concrete digital computation in Chapters 3 to 7 for cognitive scientific explanations. The main claim of this section is that for a computational explanation of cognition to be *intelligible*, irrespective of being true or false, it has to commit to a single interpretation of digital computation. The argument in support of this claim proceeds as follows.

- (P1) There are many accounts of digital computation at our disposal.
- (P2) These accounts establish at least some irreducibly different requirements for a physical system to perform digital computation and are, therefore, intensionally different.
- (P3) These accounts denote different classes of digital computing systems and are, therefore, extensionally different.
- Hence, (IC/P4) Extant accounts of digital computation are intensionally and extensionally non-equivalent, thereby rendering the term 'digital computation' ambiguous.
- (P5) A computational explanation of cognition uses the term 'digital computation'.
- (P6) If an explanation uses an ambiguous term, it is unintelligible without a commitment to a single interpretation of that term.
- Therefore, (C) A computational explanation of cognition is unintelligible without a commitment to a single interpretation of 'digital computation' as defined by a given account thereof.

The truth of the first premise (P1) is evident in the philosophical literature. Chapters 3-7 examined nine accounts of digital computation (if the Triviality account is also included). Still, there are other plausible accounts based, for example, on digital state machines or the calculation of functions.

The fifth premise (P5) is trivially true by definition, but it is also empirically grounded. Many computationalists take it for granted (cf. Anderson, 1983; Fodor, 1975; Marr, 1982; Newell & Simon, 1976; Pylyshyn, 1984; van Rooij, 2008) and so do some connectionists. In Section 8.3.1, we make a further distinction between various types of computationalism. Here we simply refer to *digital* computationalism. Extreme dynamicists reject digital computationalism (Pfeifer & Scheier, 1999; Thelen & Smith, 1994; van Gelder & Port, 1995; Wallace, Ross, Davies, & Anderson, 2007), due to the presupposition that *digital computation* is inherently representational, in the extrinsic sense (see Section 8.3.4). Still, other dynamicists do not deny that *some* aspects of cognition may be subject to a computational explanation (Beer, 2013). 'Computation' and 'computationalism' have become associated with the symbolic tradition, but they need not be limited to either symbolic computation or classicism, respectively (this is further discussed below). For the purposes of the present argument and given the focus of this book, we restrict the scope of computational explanation of cognition to *digital computationalism.*

The sixth premise (P6) calls for disambiguation when there is equivocation. The use of a term that is open to two or more interpretations is problematic, since it is a potential source of unintelligibility, or misleadingness, at the very least. To avoid ambiguity, one should commit to a single interpretation of that term. For instance, 'depression' has at least two typical interpretations. In the sentence, "The great depression started in most countries in 1929 and lasted for several years", it is clear that 'depression' means a long-term downturn in economic activity. On the other hand, in the sentence, "Long depression leads to making irrational decisions", 'depression' may mean a specific mental condition. Analogously, when one asserts that hierarchical planning or linguistic tasks, for example, are computational, one ought to commit to a single interpretation of (digital) computation. (Of course, 'digital computation' is not ambiguous in the *same* manner that 'depression' is.) Is it in virtue of executing an algorithm, formally manipulating symbols, or implementing a TM that cognitive agents engage in hierarchical planning?

Furthermore, any commitment to a particular interpretation should be consistent to avoid ambiguity. From the two sentences above concerning depression it follows that irrational decisions were made in the countries that suffered the great depression in 1929. Generally, this conclusion would only validly follow from these two sentences as premises, if 'depression' had the *same* interpretation in both cases. Otherwise, whilst this conclusion may be *plausible*, it does not *necessarily* follow. This mistake in reasoning is also known as the fallacy of equivocation. If the interpretation of a shared term changes between the premises, then the argument is invalid. To avoid this fallacy, arguments ought to comply with the requirement of a uniform interpretation. The same interpretation should be used by all premises leading to a single conclusion.

Let us now consider the second premise (P2). Clearly, some requirements for a physical system to compute implied by one account of computation *could be* reduced to other requirements implied by another. That is, there could be *some* overlap between the sets of requirements implied by different accounts. But an overlap among requirements does not imply the reduction of one requirement to another. If *all* the requirements could be reduced to a coherent minimal set of key requirements, then this would constitute a *single* account of computation. The fourth premise (IC/P4) would no longer follow from the preceding premises (P1)-(P3). However, (P2) only requires that *some* requirements do not overlap.

Whilst there is certainly some overlap among the requirements implied by the different accounts, there remain some requirements that are irreducible to others. We first consider the overlap among some of the requirements. For instance, Copeland's algorithm execution account presupposes the fourth requirement of Turing's account, that is, that the system has the capacity to follow instructions. Further, the requirement that the system has some form of memory is shared by many of the accounts examined. For example, it is specified by the second requirement of Turing's account. It is also specified by the third requirement of the PSS account and the fifth requirement of the FSM account. In a weaker form, it is specified by the fourth requirement of the mechanistic account too.

The FSM and the PSS accounts clearly share many similar requirements. One similarity is the questionable 'maximal plasticity of function' requirement advocated by Pylyshyn and the combination of the second and fourth requirements of the PSS account. These requirements, broadly construed, imply the programmability of digital computing systems. In addition, both the third requirement of the FSM account and the fourth requirement of the PSS account demand the system's capacity to handle an unbounded number of symbolic representations. Importantly, both the PSS and the FSM accounts share a *symbolic interpretation* of Turing's account, and accordingly they presuppose the first two requirements of Turing's account.

Nevertheless, despite these similarities, there are many key requirements implied by the accounts examined in this book that cannot be reduced to the other requirements. For example, the second and fourth requirements of the mechanistic account emphasise the synchronicity of components of computing mechanisms. Yet, the other accounts remain silent, for the most part, on this aspect.

Furthermore, whilst the FSM and PSS accounts take digital computing systems to engage in information processing at the symbol level, other accounts do not. For instance, Fodor and Pylyshyn claim that "conventional computers [...] process information by operating on syntactically structured expressions" (1988, p. 52). Newell and Simon's fundamental working assumption was that "the programmed computer [... is a] species belonging to the genus IPS [i.e., information processing systems]" (1972, p. 870). This bears strong resemblance to an IP account that is based on *factual* information, but other accounts, most notably, the mechanistic account, do not equate digital computation and information processing. According to the mechanistic account, digital computation does not entail information processing, because digits *can*, but *need not*, carry information.

Another example of an irreducible requirement is the situatedness of the computing system. This is the first requirement of Smith's participatory account. Although other accounts may presuppose some interaction with the environment, that can be very limited, say, in the case of offline computation, they do not specify any explicit requirement to that effect (with the exception of the PSS account).

Also, only some accounts are committed to concrete digital computation being the execution of programs, whereas others are not. The FSM and the PSS accounts are committed to digital computation being a program executed on symbolic expressions (cf. the first requirements of both the FSM and PSS accounts). Copeland's algorithm execution account, as discussed in the previous chapter, can be construed as equating digital computation with the execution of a *program*. The IIP account certainly considers a process that executes a program to be computational, but does not reduce digital computation to program execution. In like manner, the mechanistic account does not equate digital computation with program execution.

The examples considered above suffice to show that the accounts examined here establish different requirements for a physical system to digitally compute. These accounts are intensionally different, since digital computation is constituted by different properties according to each of them.

We have seen that these accounts also denote different classes of digital computing systems, and so, as stated by the third premise (P3), they are extensionally different (see Figure 8.1). Let us recap why this is the case. For one thing, both the Searle-triviality thesis and the Putnam-triviality theorem imply that every sufficiently complex system performs *every* digital computation. The Turing account can be interpreted as being either too inclusive, if it falls prey to Searle's trivialisation of digital computation, or too restrictive for excluding modern digital computers that perform multitasking. Similarly, when Smith's participatory account is judged in isolation (that is, not as an extension of another account) too many digital systems, including digital thermometers, DVD players and CRT TV sets, qualify as genuine computing systems.

Moreover, on the FSM and PSS accounts only symbolic computation qualifies as genuine computation. The PSS account denotes a narrow class of digital computing systems that only includes *universal* digital computing systems. Special-purpose computers, discrete connectionist networks and FSA, for

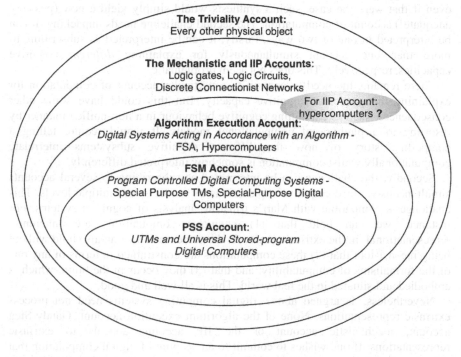

Fig. 8.1 This diagram illustrates the relations among the classes of digital computing systems denoted by six different accounts of computation. Note that the IIP account remains neutral on hypercomputation, but is otherwise equivalent to the mechanistic account.

example, are excluded from this class. According to the FSM account, logic circuits, FSA and discrete connectionist networks are classified as non-computational. On this account, the class of digital computing systems is restricted to *symbolic program-controlled* computing systems.

The algorithm execution account, the mechanistic account and the IIP account, on the other hand, do not exclude paradigmatic digital computing systems, whilst, at the same time, they are not too inclusive. The algorithm execution account includes conventional digital computers, FSA, TMs and, when it is not limited to *classical* algorithms, hypercomputers as well. But it also excludes non-computing systems, such as the solar system or digestive systems. The mechanistic account classifies more systems as computational than the algorithm execution account, namely discrete neural networks, primitive logic gates and combinational circuits. With the exception of hypercomputers, on which it remains silent, the IIP account also classifies these systems as computational.

Premises (P1)-(P3) imply the intermediate conclusion (IC/P4), and premises (P4)-(P6) entail the conclusion of the argument (C). Possible challenges to the conclusion (C) might be that some of the key requirements implied by different accounts of concrete computation could be synthesised or that one could simultaneously subscribe to two accounts or more. The first challenge may result in sidestepping the demand to commit to a single account of computation. But even if that were the case, such a synthesis would simply yield a new (possibly adequate!) account of computation. The second challenge needs unpacking. It can be interpreted in one of two ways. Firstly, it can be interpreted as subscribing to more than one account simultaneously for explaining *different* cognitive capacities, respectively. This does not seem problematic.

There remains the need to commit to a particular account of computation for explaining the particular cognitive capacity. But this could have some other consequences, such as explaining cognitive behaviour in a non-unified manner by resorting to a plethora of computational models. It would require telling a compelling story of how the different cognitive subsystems interrelate computationally whilst computation is somehow interpreted differently.

Secondly, the challenge can be interpreted as subscribing to several accounts simultaneously, since cognitive explanations by nature span multiple levels. This challenge is compatible with Marr's tripartite analysis of cognitive systems. For instance, we can hold that (1) *cognitive* computations are inherently representational, in the extrinsic sense. At the same time, we can also hold without being inconsistent that (2) these computations are constrained in terms of any one of the formalisms of computability, and that (3) they occur in the brain, which is embodied and situated in the real world. This is all well and good.

Nevertheless, as argued above, digital computing systems need not process extrinsic representations. None of the algorithm execution account, Gandy-Sieg account, mechanistic account or the IIP account appeals to extrinsic representations. If one wishes to commit to an account of digital computation that appeals to extrinsic representations, since *cognition* is representational, one ought to, first and foremost, justify *why* computation proper *is* extrinsically representational. Note that subscribing to an account of concrete computation and to a formalism of computability simultaneously does not introduce any conflict, as discussed in Chapter 1.

8.3 The Explanatory Role of Computation in Cognitive Science

There are currently considerable confusion and disarray about just how we should view computationalism, connectionism and dynamicism as explanatory frameworks in cognitive science. In this section, we examine the degree to which they are in conflict versus their degree of overlap, whether they are explanatory or merely descriptive, which levels of analysis they belong to and their explanatory posits. An important distinction that should be drawn in this context is between the conceptual issue of how computation is best characterised and the empirical issue of how cognition is best explained.

On the one hand, as regards the empirical issue, computationalism, connectionism and dynamicism are not in competition. A single cognitive system might be correctly modelled within each one of these paradigms. It is also even possible that all of them include some form of computation simpliciter (insofar as dynamicists are willing to accept it as being explanatorily relevant). So, in this sense, the three paradigms may explanatorily coexist, if within each paradigm the *same* cognitive system is modelled in *different* ways. Still, this implies that dynamicism also offers *models* of the cognitive phenomena in question.

On the other hand, viewed from a different perspective, they are in conflict. Either the bulk of cognitive phenomena are best explained symbolically or they are not. And if they are best explained symbolically, then a particular form of digital computation can indeed be central. This was the crux of the classicist-connectionist debate in the late 80s and throughout the 90s. Further, either the bulk of cognitive phenomena are best explained in a disembodied/non-embedded manner or they are not. Here enters the extreme dynamicist to the debate denouncing the computationalist and connectionist explanatory paradigms.

'Digital computation' is an ambiguous concept and is the source of an ongoing conflict among the central paradigms in cognitive science. This conflict stems from an equivocation on the notion of computation simpliciter. Computation is invoked differently by classicism, computationalism – broadly construed, connectionism and computational neuroscience with varying degrees of success. It is not just the dichotomy between analogue and digital computation that is the basis for this equivocation, but also the diversity of extant interpretations of *digital* computation. Analogue computation has received even less attention in the literature, and much like its digital counterpart, it remains equivocal.

Two main arguments are presented in this section. First, a blanket dismissal of the key role computation plays in cognitive science is unwarranted. For 'computation' is an ambiguous concept and is invoked differently across a range of research programs in cognitive science. Whilst some accounts of concrete *digital* computation proper are untenable, others remain plausible and have important implications for computational explanatory paradigms that are underpinned by them.

Second, the thesis that computationalism, connectionism and dynamicism are mutually exclusive is wrong. For computationalism can be narrowly construed as classicism, but also more broadly as digital computationalism, generic computationalism or even pancomputationalism (or also as *analogue* computationalism). Further, connectionism is compatible with generic computationalism, since it may be classified as either digital or analogue computationalism, depending on the type of connectionist networks used. Digital computationalism and connectionism make available mechanistic models of the cognitive phenomena in question. But dynamicism proper is not on a par with either connectionism or digital computationalism, as it does not necessarily offer a mechanistic explanation of the cognitive phenomenon.

8.3.1 Computationalism and Computation Simpliciter

8.3.1.1 Classicism and Symbolic Computation

The classicist thesis is that cognition is symbolic computation. Pylyshyn claims that the idea that "certain behavioral regularities can be attributed to different representations (some of which are called beliefs [...]) and to symbol-manipulating processes operating over these representations" is fundamental to cognitive science (1999, p. 10). Similarly, Haugeland claimed that "thinking and computing are radically the same" (1985, p. 2) and that "an intelligent system *must contain* some computational subsystems [...] to carry out [...] internal manipulations" (1985, p. 113, italics added).

Classicists are committed to the idea that cognitive capacities are underpinned by mental representations. They insist that the combinatorial structure and compositionality of mental representations are critical for our cognitive capacities (Fodor & Pylyshyn, 1988, pp. 17–18). Fodor, for example, claims that cognitive processes are operations defined on syntactically structured mental representations, much like sentences in natural language (1981). Pylyshyn adds that "[w]hat makes it possible for [...] intelligent organisms to behave in a way that is correctly characterised in terms of what they represent (say, beliefs and goals) is that representations are *encoded* in a system of physically instantiated symbolic codes" (1999, p. 5, italics original).

Classicism then is a narrow conception of digital computationalism. It is committed to a multi-level symbolic model of cognition. Physical symbol systems, for example, are describable at least at two levels: the symbol level and the physical level. Symbol structures and operators on these structures, at the symbol or program level, are realisable in physical mechanisms (Newell, 1980a, p. 156). Physical symbol systems are also describable at the knowledge level, which is characterised by knowledge as a potential for generating action relative to the goals of the system (Newell, 1980b). In a manner akin to Marr's tripartite analysis[1], Pylyshyn also offers a tripartite decomposition of cognitive systems. At the top/semantic level, knowledge and goals as well as certain behaviours of the cognitive system are attributed to different representations and the processes operating on them, respectively. At the middle/symbol level, symbolic expressions encode the semantic content of the system's knowledge and goals. At the bottom/physical level, representation-governed behaviour of the entire system is implemented by some biological substrate (Pylyshyn, 1993, 1999, pp. 7–11).

It is easy to see then why classicists invoke computation-theoretic language for explaining cognition. In the light of the above, the great flexibility of programmable digital computers makes them ideal models of cognitive agents performing complex tasks in virtue of language-processing-like operations. By endorsing either the FSM or PSS accounts of computation, classicists can easily appeal to known computational architectures and related tools to explain cognitive phenomena.

8.3.1.2 Broad Construals of Computationalism and Digital Computation

How broadly can we construe computationalism? The classical dichotomy of computation simpliciter is between digital and analogue computation. Even if we took computation to be just *digital* computation, we would still be left with many versions of digital computationalism depending on the particular interpretation adopted of computation. Broad *digital computationalism* is certainly more encompassing than classicism, which posits a narrow class of digital computing systems. A classicist, who subscribes to the FSM account, takes physical computing systems to be program-controlled digital computers. Her fellow classicist, who subscribes to the PSS account, takes physical computing systems to be programmable stored-program digital computers.

Importantly, different accounts of digital computation can give rise to different versions of broad digital computationalism. For example, on the Triviality account, every sufficiently complex physical system computes every Turing-computable function. This view inevitably leads to *strong digital pancomputationalism*, since rocks, chairs, paper clips, oranges, humans and the physical universe – all digitally compute. This version of digital pancomputationalism is hardly illuminating. More precisely, it is an anti-realist version of pancomputationalism. It does not tell us, for example, that the universe has a particular structure, but it is rather aimed at arguing against the idea that cognition is computational in any substantial sense (Dodig-Crnkovic & Müller, 2011, p. 154).

However, we need not go so far as to promote digital pancomputationalism, to be able to endorse a digital computationalist thesis that is broader than classicism. Subscribing to some of the other accounts of concrete digital computation, which are not limited to symbolic computation and do not appeal to extrinsic

---

[1] The semantic level, for example, is sometimes equated with Marr's top/computational level, but it should not be. Marr's top level characterises the function computed by the cognitive system. The computation can, but need not, involve the assignment of semantic contents.

representations, leads to a broader, more intelligible, digital computationalist thesis.

Consider first the resulting digital computationalist thesis based on the mechanistic account. Call it the MDC thesis. The MDC thesis is broader than the classicist thesis. Unlike the FSM and PSS accounts of computation, it is not restricted only to symbolic computation. The mechanistic account is adequate for explaining computation proper and offers specific criteria for classifying a physical system as computational. It depends on a conceptual decomposition of the computing mechanism concerned into its fundamental components, functions and organisation. In the right context where any relevant provisos about the cognitive system concerned are clearly specified, the MDC thesis is empirically testable. Whether a physical cognitive system is a nontrivial computing mechanism can be verified empirically. Because the mechanistic account encompasses many possible computing systems, including basic logic gates, special-purpose computers, programmable stored-program systems and discrete neural networks, the MDC thesis is not committed a priori to cognition being any one particular type of computing system. The MDC thesis is, at least in principle, broader than the classicist thesis.

Some of the other accounts of computation examined in this book can also give rise to computationalist theses that are broader than the classicist thesis. The IIP account too is adequate for explaining computation proper and does not appeal to extrinsic representations. It offers specific criteria for classifying a physical system as computational in terms of data and instructional information processing that is done either conditionally or unconditionally. Combined with any clearly specified supporting conditions and provisos about the postulated cognitive system, the resulting IIP thesis of cognition can be empirically tested. In a similar manner to the resulting MDC thesis, an IIP thesis of cognition is not committed a priori to cognition being any one particular type of computing system.

Copeland's algorithmic execution account can too give rise to a computationalist thesis that is broader than the classicist thesis. For it does not appeal to extrinsic representations nor is it limited to symbolic computation. However, its inadequacy for explaining computation proper has implications for a computational explanation of cognition. A cognitive system can only be deemed computational if it executes a program. But, if that is true, then a connectionist explanation of cognition, for example, cannot be underpinned by Copeland's algorithmic execution account. (In the next section, it is argued that some connectionist networks perform digital computation).

Computationalism may be further extended beyond *digital* computationalism. If 'computation' is taken as *generic computation*, then we get the broadest version of computationalism, which is still *not* digital *pan*computationalism. Generic computation includes digital computation, analogue computation, quantum computation and neural computation. It is characterised as the processing of medium-independent vehicles according to rules allowing for the processing of continuous variables, strings of digits or neuronal spike trains (Piccinini & Scarantino, 2011, pp. 10–13). *Generic computationalism* is the thesis that cognition is computation in a generic sense. But it does not amount to pancomputationalism, since not *every* physical process is computational. It is a plausible, albeit weak, explanatory framework. It is still falsifiable, as it could turn out, for example, that cognitive capacities depended inherently on some medium-*dependent* properties.

In sum, computationalism should not be identified with classicism. The latter is but one digital computationalist alternative positing a narrow class of digital computing systems as candidate models of cognition. How broadly computationalism should be construed depends on the particular interpretation adopted of computation.

8.3.2 *Connectionism and Sub-symbolic Computation*

The connectionist thesis is that cognition is sub-symbolic computation. Accordingly, cognition should be explained by connectionist network activity, in a more generic sense than the association between stimuli and responses. Modern connectionists argue that these networks perform sub-symbolic computation (Chalmers, 1992; MacLennan, 2001; Rumelhart & McClelland, 1986; Smolensky & Legendre, 2006; Smolensky, 1988). On this view, it is implicit that connectionist networks broadly are capable of computations that are not limited to discrete manipulations of symbolic representations. In a sharp contrast to classicism, most connectionists reject the claim that a language of thought is required for an adequate explanation of cognition. This makes the tension between connectionism so construed and classicism obvious. Some have explicitly advanced the thesis that connectionist networks are analogue computing systems (Diederich, 1990; O'Brien & Opie, 2006; O'Brien, 1999). Others have restricted the classification of connectionist networks as analogue computing systems to a specific type of networks, primarily those that process real-valued quantities (Kremer, 2007; Siegelmann, 1999).

According to Gerard O'Brien and Jon Opie, connectionism is grounded in analogue computation, for connectionist networks "compute by exploiting relations of structural resemblance between their connection weights and their target domains" (2006, p. 41). On their view, "[a]nalog computers are systems whose behaviour is driven [...] by semantically 'active' analog representations that physically or structurally resemble what they represent" (O'Brien & Opie, 2006, p. 33). It follows, by their lights, that connectionist networks are analogue computing systems. The representational vehicle invoked in a connectionist analysis is based on a structural isomorphism between the network's activation patterns and the task domain. This isomorphism renders the shape of the activation landscape *semantically* significant (O'Brien & Opie, 2006, pp. 32–34; O'Brien, 1999).

If they are right, then connectionism is a variant of *analogue computationalism*, that is, the thesis that cognition is analogue computation. But, as mentioned above, the notion of analogue computation is also equivocal. According to O'Brien and Opie, analogue computation is defined over analogue representations. However, when Hava Siegelmann, for example, invokes this notion, she refers to computation performed by a very specific type of recurrent neural networks that perform operations on real variables (1999).

Nevertheless, the most precise characterisation of analogue computation may be attributed to the Shannon-Pour-El Thesis. According to this thesis, the outputs of general-purpose analogue computers correspond exactly to differential algebraic functions. Therefore, there exists a universal analogue computer that using just a handful of integrators can compute, to some arbitrary degree of approximation, any possible continuous function (Rubel, 1985, pp. 75–76). The main point is that analogue computation is a continuous change of real variables over time.

Yet, there remain the questions whether analogue computation has to be defined over representations and whether connectionist networks are rightly classified as analogue computing systems. Analogue computing systems and their processing units have the function of transforming an input real variable into an output real variable that stands in some specific functional relation to the input variable. Their operations can, but need not, be understood in terms of analogue representations (Piccinini, 2008a, p. 48). There are certainly good reasons to classify *discrete* connectionist networks, which process binary-valued or integer-valued quantities, as digital computing systems. That would certainly be the case, if we adopted, say, either the mechanistic account or the IIP account.

Some connectionist networks perform digital computation, whilst others perform analogue computation. The idea of discrete binary networks goes back to the seminal 1943 paper by Warren McCulloch and Walter Pitts. On their model, each neurone was modelled as a linear threshold element with a binary output. This was the first model of a discrete connectionist network exhibiting all-or-none firing patterns. When both the inputs and the outputs of such connectionist networks are binary the result is a Boolean circuit (Siu et al., 1995, pp. 1–2). Since McCulloch-Pitts networks can be used to build digital computers, these, and similar discrete, connectionist networks are best classified as digital computing systems. Otherwise, if McCulloch-Pitts networks were classified as analogue computing systems, then digital computers would be analogue too.

Furthermore, we can distinguish between two types of connectionist networks based on their dynamics. According to the mechanistic account and the IIP account, the first type of networks takes strings of digits – or discrete data on the IIP account – as inputs and outputs, has discrete dynamics and does not change its structure over time. The second type of networks takes strings of digits as inputs and outputs, but has continuous dynamics or changes over time (Piccinini, 2008b). Whilst only the first type belongs to the class of *classical* digital computing systems, on the mechanistic account, both these types of connectionist networks perform digital computation.

Yet, there exists another class of connectionist networks that process continuous real-valued quantities and *do not* perform digital computation. These networks turn their input into their output in virtue of their continuous dynamics and do not compute by manipulating strings of digits (Piccinini, 2008b, p. 319). Continuous variables are not strings of digits and this, on the mechanistic account, suffices to rule out such networks as non-computing systems in the sense of computation employed in computer science. Nevertheless, these connectionist networks can, arguably, be classified as analogue computing systems, for they have five properties that are typical in analogue computers. First, the network's operations take a continuous range of values over time. Second, its physical dynamics are governed by operations on real variables. Third, the functional relation between inputs and outputs of the network is describable by a set of differential equations. Four, the network's inputs and outputs are distinguished from one another up to a limited degree of precision. Lastly, the network is subject to varying levels of noise (Fresco, 2010, pp. 24–25).

Granted that connectionist networks compute, connectionism is, at the very least, a subclass of generic computationalism. If *discrete* connectionist networks are *sufficient* for explaining cognition, then connectionism does not just overlap with, but is a *subclass* of digital computationalism. If *continuous* connectionist networks are *sufficient* for explaining cognition, then connectionism is a subclass of analogue computationalism. However, if the full range of connectionist networks is required for explaining cognitive phenomena, then connectionism is a subclass of generic computationalism. At any rate, on the preceding analysis, none of these three options has to presuppose extrinsic representations for connectionist computation.

8.3.3 Computational Neuroscience and Neural Computation

Already in the 90s, but particularly in the past decade, computational modelling of cognition has become an active area in neuroscience in an attempt to disclose how neurones give rise to cognitive functions. This research program now wears the title computational neuroscience and employs a broad range of techniques using some tools from the domain of computer science.

Computational neuroscience should not be identified with connectionism. The latter typically refers to models based on behavioural data, whereas the former refers to models based on *both* behavioural and neuroscientific data. Besides, the backpropagation method, which is often used to train connectionist networks, depends on neurones being able to relay signals bi-directionally. Dendrites and axons, on the other hand, acting as input and output channels to and from brain neurones typically allow nerve impulses to travel in one direction only. And whilst individual connectionist neurones are homogenous, brain neurones are physiologically specialised.

Computational neuroscience downplays the explanatory role of the standard digital computer metaphor and connectionist networks in cognitive science. Computational cognitive science attempts a fairly close integration of psychological, neurophysiological and neurobiological data and theories of cognition (Boden, 2008). Most existing connectionist networks are hugely different from the anatomy of the brain. Connectionist neurones are computationally far too simple when compared with real neurones, though some attempts have been made to model brain neurones more faithfully (cf. the discussion about models that do not impose the simplification or homogenisation of the computational units in (Maass & Markram, 2004)). Some degree of simplification is clearly needed to make *any* model viable, since models, by definition, abstract from some of the particulars of the modelled system.

The alleged incompatibility of the standard computationalist analysis with the organisation of the nervous system has been pointed out by reference to Marr's tripartite analysis (Churchland & Sejnowski, 1992, pp. 18–19). On Marr's analysis, the top-level competence function can be examined independently of understanding the algorithm that is performed in the brain and similarly the problem of discovering the algorithm at work is independent of its underlying physical realisation. This top down approach makes neurobiological facts about the nervous system supposedly less pertinent, since they are just details at the implementation level. Yet, further research in computational neuroscience has suggested that knowledge of the brain architecture plays a vital role in understanding those "algorithms that have a reasonable shot at explaining how in fact neurons do the job" (Churchland & Sejnowski, 1992, p. 19).

Unlike digital computationalism, computational neuroscience studies cognition in a bottom-up approach, whilst being informed by top-down theories. Research from neuropsychology, neuroethology and psychophysics provides the details about the relevant lower level neural mechanisms. But lower level research remains incomplete in the absence of top-level analyses of the very cognitive capacity, whose mechanisms are studied at the lower level. Neuroscientific research can profit, for instance, from abstract discoveries in computability theories and discoveries in the construction of physical computing systems (Churchland & Sejnowski, 1992, pp. 11–12). Neuroscience attempts to do more than "merely reproduc[e ...] a function of the brain (such as playing chess)", whereas this may be sufficient in AI research (Eliasmith & Anderson, 2003, p. 1).

Nevertheless, as its name suggests, computational neuroscience is committed to the view that the brain *is* an implemented computing system (Churchland, Koch, & Sejnowski, 1990; Churchland & Sejnowski, 1992; Dayan & Abbott, 2001; Eliasmith & Anderson, 2003; Trappenberg, 2010). But it is not committed to cognition being either symbolic computation or sub-symbolic computation. Brain neurones are taken to be computational units that process information to solve complex tasks, such as perception. What neuroscientists take 'computation' or 'neural computation' to be is another matter entirely. One approach is to agree that whilst there is no precise definition of neural computation performed by the brain, it is broader than just digital computation. Some neuroscientists take a computing system to be one whose physical states can be described as representing states of some other systems, where transitions between states are operations on representations. And neural computation, in particular, is taken to be the encoding and decoding of neuronal spike trains (Eliasmith, 2007, pp. 326–327).

Arguably, so-called neural computation may be a *sui generis* type of computation. This has been a recent thesis of some researchers (Piccinini & Bahar, 2013). They argue that neuroscientific evidence shows that typical neural signals, such as spike rates, are continuous. On the other, these neural signals are constituted by spikes that are discrete elements. Whilst this thesis is not uncontentious, it is compatible with other characterisations of neural computation in neuroscience according to which neural computation is neither exclusively digital nor analogue computation (Churchland et al., 1990, pp. 47–50; Eliasmith, 2007, pp. 326–327; Poggio & Koch, 1985). Others have proposed *natural computation* as an alternative notion of computation that is more suitable for describing the behaviour of biological systems (Hamann & Wörn, 2007; MacLennan, 2004). For one thing, neural activity has many sources of noise that may distort the underlying computation. This suggests that, unlike digital computation, natural computation itself is noisy and sometimes distorted (MacLennan, 2004, p. 129). The claim that neural computation, as it is used in computational neuroscience, is a *sui generis* type of computation needs unpacking and will await exploration in future research.

8.3.4 Extreme Dynamicism and the Non-computational Shift

Various "anti-representationalist" approaches are included under this heading starting with "radical" dynamicism (e.g., Thelen & Smith, 1994; van Gelder & Port, 1995) through embodied and embedded dynamicism (e.g., Pfeifer & Scheier, 1999) to the enactivist approach (e.g., Thompson, 2007; Varela, Thompson, & Rosch, 1991). Whilst there are important differences amongst these approaches and grouping them together certainly does them an injustice by blurring those differences, they all share a similar trait. They all reject representation and computation as being key to understanding cognition.[2] Instead, according to this new "post-cognitivist" paradigm, cognition is not computational (Wallace et al., 2007, p. 26). This new paradigm distances itself from both computationalism and connectionism by broadening its research focus on the brain and including the body and its relationship to the "outside" world. The purpose of the brief exposition here is to reveal any misconceptions about computation and so specific details about the different approaches are omitted.[3]

An underlying claim of the extreme dynamicist approaches is that cognition is *not* computational. Advocates of these approaches maintain that it is time for cognitive science to embrace a non-computational paradigm. In the early 90s, Rodney Brooks designed the *mobots*, which were robots capable of functioning in a messy and unpredictable environment. He claimed that these robots "do not have traditional AI representations [...] which have any semantics that can be attached to them" (Brooks, 1991, p. 149). Brooks seems to reject the processing of extrinsic representations as being essential to the mobots' computation. "[T]here need be no explicit representation of either the world or the intentions of the system" (Brooks,

---

[2] The label 'extreme dynamicism' is used to alert the reader that in some sense, every cognitive scientist is, by definition, a dynamicist. For there seems to be a consensus that cognition is a dynamical phenomenon, and as such it requires some application of dynamical systems theory. For clarity, the label 'extreme dynamicism' is chosen to denote the anti-computationalist position.

[3] To be sure, these different approaches are logically autonomous. One can subscribe to any particular approach without necessarily subscribing to the others. For a nice discussion on the history and differences amongst those approaches see, for example, Evan Thompson (2007, pp. 3–15).

1991, p. 149). "Radical" dynamicists, Tim van Gelder and Robert Port, argued that "[t]he cognitive system is not a discrete sequential manipulator of static representational structures" (1995, p. 3). Similarly, embodied dynamicists, Rolf Pfeiffer and Christian Scheier, criticised the "analogy between human thinking and processes running in a computer, that is, information processing as the manipulation of symbols" (1999, p. 47).

Researchers endorsing one or more of these approaches have rejected the cognitivist paradigm, which gives rise to some form of a "Cartesian theater" (Spivey, 2007, p. 313) and relies on a metaphor of the "mind as a computer" (Spivey, 2007, p. 29). Still, the common interpretation of a computer is as a serial digital system (Froese, 2011, p. 118; Spivey, 2007; Wallace et al., 2007, p. 10) that performs information processing on representations (Thompson, 2007, p. 186). For the extreme dynamicist, representation is not a mandatory concept for explaining cognitive phenomena, which are seen as the simultaneous, mutually influencing unfolding of complex temporal structures. The digital computationalist, on the other hand, supposedly explains cognitive phenomena as simple transformations of static representations (Thelen & Smith, 1994, pp. 164–165; van Gelder, 1998, pp. 621–622).

It seems then that advocates of the various extreme dynamical approaches share a common (mis)conception of computation. This conception leads them to reject computational research programs in cognitive science. Rather than relying on computer science as the foundation for traditional cognitive science, they promote dynamical systems theory as the foundation for an *alternative* cognitive science. For dynamical systems theory provides a general mathematical theory, which supposedly is *already* the standard language of the natural sciences, and it allows us to do better justice than computability theory to the continuous temporal changes of cognitive phenomena at multiple timescales (Froese, 2011).

Extreme dynamicists take computation to be a serially digital process that is carried out over extrinsic representations. Yet, GMs, cellular automata and discrete connectionist networks – all perform *parallel* digital computation and violate this narrow characterisation. As well, computational neuroscience invokes the notion of neural computation that is, possibly, different from digital computation proper. Further, the assertion that digital computation is carried out over extrinsic representation is unsupported. We have seen in previous chapters that there can be accounts of concrete digital computation that need not appeal to extrinsic representations. Symbolic computation, which does appeal to this type of representation, is merely a narrow class of digital computation. So, the dynamicist rejection of the "processing of representations" as the basis of an adequate model of cognition can only apply to computationalist models based on, say, the FSM and PSS accounts. However, broad digital computationalism is not susceptible to a similar criticism on these grounds.

Extreme dynamicism is advanced as a non-computational more biologically plausible explanatory framework of cognition. Nevertheless, it is not obvious why this is the case, as extreme dynamicists tend to ignore the practical details of the underlying mechanisms of the cognitive systems in question. That brings us to the next section.

8.3.5 Mechanistic versus Non-mechanistic Explanatory Frameworks

Before turning to evaluate whether dynamicism, connectionism and digital computationalism should be viewed as either competing or complementary explanatory frameworks, let us briefly examine the main aspects of mechanistic explanations. Mechanisms typically have four characteristics: being phenomenal, being componential, being causal and having a clear organisation. First, they are phenomenal[4] in the sense that they perform tasks. The phenomenon is explained by appealing to the tasks performed as a whole and it partially determines the boundaries of the mechanism. Second, all mechanisms have at least two components. The components of a mechanism are those that are relevant to the explanandum. Third, these components are causally interrelated, that is, they interact with one another. Four, the spatial organisation of the components, in terms of their locations, shapes, orientations, etc., as well as their temporal organisation, in terms of the order, rates, and durations of the activities in the mechanism, play a key role in generating the phenomenon (Craver & Bechtel, 2006, pp. 469–470).

A mechanistic explanation requires isolating some aspect of the phenomenon to be explained and positing the mechanism that is capable of producing this phenomenon. It is achieved by virtue of identifying the relevant subcomponents of the mechanism and the corresponding activities (i.e., localisation) that are organised in the right way so as to produce the phenomenon in question. The localisation of the relevant components and corresponding activities is accomplished by means of *structural* and *functional* decomposition, respectively. A structural decomposition begins by breaking the mechanism apart into subcomponents and then investigating what they do. A functional decomposition is accomplished by analysing the phenomenon into activities that, when properly organised, exhibit the phenomenon concerned. For example, the chemical process of fermentation may be decomposed into a set of more basic chemical reactions, including oxidation and phosphorylation (Craver & Bechtel, 2006, p. 473).

In the context of mechanistic explanations, a distinction is typically made between mechanistic *sketches*, or *schemata*, and *complete* mechanistic models. A mechanistic sketch is a functional analysis in which some structural details of a mechanistic explanation are excluded. But once the omitted details are filled in, the functional analysis becomes a full-blown mechanistic explanation. A *complete* mechanistic model, on the other hand, identifies the functional properties of the components and has to respect constraints imposed by those components. It also does not leave out any crucial gaps regarding how the mechanism works (Piccinini & Craver, 2011). With this brief exposition in mind, let us return to examine the relation among dynamicism, connectionism and digital computationalism as explanatory frameworks.

---

[4] This term is borrowed from (Craver & Bechtel, 2006, p. 469) in a manner that is unrelated to *phenomenology*.

Recently, some devoted dynamicists have argued that good dynamical accounts of cognitive phenomena are genuinely explanatory and not merely *descriptive* (Stepp et al., 2011). This defence was invoked in response to the contemporary mechanistic philosophy of science that allegedly excludes dynamicist explanations of cognition (Bechtel, 2009; Machamer, 2004; Piccinini & Craver, 2011). According to the defenders, the reason for this exclusion results from either a theoretical commitment to computational explanations or a normative commitment to a mechanistic philosophy of science. Instead of proposing a complete mechanistic explanation, dynamical explanations seek to model cognitive phenomena by identifying higher-level laws (or law-like principles). Dynamical explanations, so it is argued, capture the temporal change of the phenomenon in question by a set of differential equations (Stepp et al., 2011, p. 432).

Other authors have argued that some dynamical explanations *are* mechanistic. Arguably the fact that dynamical explanations use mathematical tools and concepts of dynamical systems theory does not entail that these explanations are non-mechanistic. On this view, extreme dynamicism can also be used sometimes to describe cognitive *mechanisms*. Carlos Zednik offers two examples that supposedly show that dynamical models and dynamical analyses can in themselves mechanistic throughout (2011).

The first example is the infant perseverative reaching model by Esther Thelen and colleagues based on Jean Piaget's classic A-not-B task. What Zednik identifies as most significant for his claim is a tripartite analysis of an input vector, which partakes in this dynamical explanation, into a task input, a trial-specific input and a memory trace, which captures the influence of prior trials. The individual contributions of the task input, trial-specific input and memory trace can be supposedly construed as the posited component operations of a mechanism for goal-directed reaching. Expressed as variables, these operations are linked in a dynamical equation that captures their role in this mechanism (Zednik, 2011, pp. 248–249).

The second example is Randall Beer's dynamical explanation of perceptual categorisation in a simulated brain-body-environment system. The simulated system consists of a single minimally cognitive agent that is equipped with a 14-neurone continuous-time recurrent connectionist network "brain". The system is situated in a simple two-dimensional environment that features a single circular or diamond-shaped object. The object falls vertically toward the agent in the course of the trial, and the agent responds by moving horizontally to catch circles and avoid diamonds, thereby performing a categorical discrimination. By Zednik's lights, Beer's dynamical explanation features a dynamical analysis that describes the activity of two components, the embodied "brain" and the environment (Zednik, 2011, pp. 250–252).

Nevertheless, these two examples do not show that extreme dynamicism necessarily offers a mechanistic explanation. Rather, they show that dynamicism is *compatible* with mechanistic cognitive models. Zednik argues that Beer's dynamical analysis relies on the mechanistic heuristic of structural decomposition to identify two components, the embodied brain and the environment. The operations associated with each of the components are described by a detailed dynamical analysis. By doing so, Zednik puts a foot on a slippery slope. For once we allow such a simple structural decomposition, any dynamicist explanation that describes the interaction between a cognitive agent and the environment is supposedly mechanistic. Beer's model is mechanistic, but only because it includes a recurrent connectionist network that models some part of the brain. The dynamical analysis in this case complements the connectionist network model used.

The infant perseverative reaching model also does not support the claim that some dynamical models *are* mechanistic proper. Zednik implies that this model can be considered a relatively abstract mechanistic sketch, which leaves more than enough room for elaborating the possible neuroanatomical components giving rise to the goal-directed reaching phenomenon. At best, this model offers a functional decomposition of low-level processes of perception and action (Zednik, 2011, pp. 249–250). Even if it were classified as an "abstract mechanistic sketch", it would be at the "very incomplete" end of the spectrum. For, it lacks any structural decomposition of the underlying relevant components. Absent the identification of the participating components, any possible causal relations among them cannot be specified. As an incomplete mechanistic sketch, this model indeed invites a future development of a mechanistic explanation. Yet, there remains a big gap to be filled, as the model has to identify the causal structure of the system in question (Piccinini & Craver, 2011, p. 292).

A dynamical explanation should not be misconceived as an *alternative* to mechanistic explanations. The Hodgkin-Huxley model of spike generation is arguably a good example of a genuine explanation. Still, it simultaneously offers a dynamical description (comprising a set of coupled differential equations to describe the dynamics of the membrane action potential) and a mechanistic one (describing how ion channels and related activities are organised to generate action potentials). These differential equations helped guide the search for the underlying components of the responsible mechanism (Kaplan & Bechtel, 2011, p. 439).

It is the non-mechanistic dynamical approach that offers a genuinely different *kind* of cognitive science. But such an approach is not on a par with either connectionism or digital computationalism. Whether this approach is truly explanatory or merely descriptive remains contentious. The burden of proof is on the extreme dynamicist to show how her approach is explanatory in the absence of a mechanistic description. Predictions based on law-like regularities alone are at best incomplete explanations.

Whilst digital computationalism and connectionism also make available mechanistic *models* of cognitive architectures, dynamicism proper offers a mathematical *formalism* describing the evolution of complex physical systems over time. Classicism and connectionism, for instance, may be competing for the same prize. But not dynamicism, as it provides a completely different type of epistemological analysis with a different purpose than the modelling one served by the other two. Whether cognition turns out to be a programmable digital computing system, a continuous recurrent network, both or neither, has no critical implications for dynamicism. If we endorsed the view defended by Zednik, then some dynamical analyses might be considered incomplete mechanistic sketches. Still, as incomplete sketches they have to be elaborated by means of structural decomposition. Typically, connectionist network models are used to complement dynamical analyses in an attempt to identify the relevant subcomponents giving rise to the cognitive phenomenon.

Nevertheless, dynamicism proper is a non-mechanistic explanatory framework. It typically explains cognitive phenomena in one of three ways: by using idealised descriptions, by identifying a small number of variables or by using connectionist network models (Thagard, 2005, pp. 200–203). When not all influencing variables can be identified and the equations cannot be spelled out, the cognitive phenomenon is described using dynamic systems terminology to varying degrees of idealisation. The changes of the complex system are described in terms of changes in state space and phase transitions. In other cases, where a small number of variables can be identified, dynamicism provides a mathematical description of the overall system state and its predicted changes under certain conditions. When connectionist network models are employed, dynamicism offers a mathematical framework for analysing the workings of these models, thereby revealing the overlap between dynamicism and connectionism.

"Radical" dynamicism, in particular, rejects the need to identify the various parts comprising the overall cognitive system and their organisation in a manner that contributes to the overall system activity. It thereby violates the decomposition principle of mechanistic explanations (Bechtel, 1998a, 1998b). "Radical" dynamicists seek to identify the laws governing the "highest level relevant to an explanation of cognitive performances, whatever that may be" (van Gelder, 1998, p. 619).

By contrast, connectionism and digital computationalism provide a mechanistic explanation of cognition. Models of cognitive architecture are available within each of these paradigms, though classicists downplay the importance of the particular physical implementation. There is certainly little reason to insist on a narrow view of computationalism as the basis for computational cognitive science. But if we adopted a broad view of digital computationalism, say, one that is based on the mechanistic account or the IIP account, then the result could be a mechanistic explanation of cognition.

Some authors have recently rejected the view that digital computationalism and connectionism are mechanistic cognitive models. For instance, Daniel Weiskopf argues that though they have some features in common with mechanistic models proper, they crucially differ in the manner in which they relate to the modelled cognitive system (2011, p. 314). By his lights, cognitive models are causally structured, componentially organised, and semantically interpretable. Such cognitive models can be specified at different levels of analysis in a similar manner to full-blown mechanistic models (Weiskopf, 2011, p. 327). However, the objection continues, cognitive models need not be mechanistic to be genuinely explanatory. For there need not always be a one-to-one correspondence of every component of the cognitive model to some *real* entity in the modelled system.

Whilst some cognitive models may be genuinely explanatory, if a one-to-one correspondence does not obtain, then, by the mechanistic standards, they are supposedly inadequate. In some cognitive models, which offer functional layered analyses, what matters is that there is some stable pattern of organisation in the brain that carries out the appropriate processes assigned to each layer of analysis, and has the appropriate sort of causal organisation. For example, there could be a correspondence to a whole set of resources possessed by neural regions, rather than, say, individual neurones. Yet, if a simple correspondence among components of the cognitive model and some neural entities in the brain does not obtain, then the model is at best incomplete, and at worst false (Weiskopf, 2011, pp. 329–330).

The gist of the objection is that digital computationalist and connectionist models are both *componential* and *causal*, but they need not be *mechanistic*. For these models often posit elements that do not straightforwardly map onto localised parts of the modelled cognitive system (Weiskopf, 2011, p. 332). However, as Weiskopf acknowledges, this objection may be countered by distinguishing between mechanistic sketches and complete mechanistic models.

Connectionism typically offers cognitive explanations in terms of connectionist networks that need not correspond to networks of real brain neurones and synapses. A single artificial connectionist neurone may correspond to a single *region* in the brain instead. Connectionist networks implement a task analysis without necessarily decomposing their overall operation into intelligible subtasks performed by individual components (i.e., a connectionist neurone) that correspond to either individual brain neurones or regions. Connectionist modellers, typically, build their networks as mechanistic models, yet they cannot give a complete mechanistic analysis of the microfeatures and microactivities that result from its adaptive weight changes during learning (Bechtel & Abrahamsen, 2002, p. 268).

Moreover, connectionist networks explain cognitive phenomena without employing localisation and decomposition. The overall performance of the network is typically not decomposable into intelligible subtasks. Instead, such networks emphasise dynamic behaviour that corresponds to the cognitive activity to be explained without the subcomponents of the system performing recognisable subtasks of the overall task. Each one of these subtasks is distributed across the layers of network and cannot be straightforwardly localised in any individual neurone. In the absence of explicit identifiable rules, connectionist networks have structures that are found in the networks' connections (Bechtel & Richardson, 2010, pp. 217, 222–223). Yet, these networks are, at the very least, mechanistic sketches.

Why are digital computationalist models mechanistic? In digital computationalist models, functional decomposition is, typically, accomplished by modelling the target cognitive phenomenon through a series of algorithmic operations. In principle, it is easier in non-connectionist computational models to localise individual operations in corresponding components, due to the nature of these models. Nontrivial digital computing systems are generally driven by a set of rules or instructions (cf. the mechanistic account, the IIP account and the algorithm execution account, but even the FSM and PSS accounts). Data (or symbols, on the classicist view) are manipulated by either hard- or soft-programmed instructions. For the purposes of classifying digital computationalist models as mechanistic, the instructions for the manipulation of data (or symbols) embody an attempt to account for the performance of the modelled cognitive system by way of decomposing the overall task into simpler subtasks.

Consider, for example, Marr's computational model of vision and John Anderson's ACT* production model. At the top/computational level, Marr's analysis specifies what is being computed and why. At the middle/algorithmic level, the visual system is specified by means of the representations being used as well as the algorithm for transforming inputs to outputs. This level provides an explanation of the structure of visual processes. The bottom/implementation level specifies the physical realisation of the representations and algorithm (Marr, 1982). Marr's tripartite model attempted to identify individual operations with specific neuroanatomical structures (i.e., localising the detection of zero-crossings in cortical simple cells). Anderson's ACT* production model analyses cognitive memory function while also providing a cognitive architecture. This model consists of three components: working memory, declarative – or explicit – memory and production – or implicit – memory. This model exhibits the performance of an action as loop of encoding, into working memory, match against a rule in production memory, and execution in the working memory (Anderson, 1983).

These two computationalist models assume that the modelled cognitive activity is decomposable into a set of operations, each of which is governed by a set of instructions operating on representations (Bechtel & Richardson, 2010, pp. 211–212). If a computationalist model also specifies how the relevant components are realised by neuroanatomical structures, then, by the standards of mechanistic explanation, it is a complete mechanistic model. Yet, such a direct localisation is not always practically feasible.

Of course, not all advocates of dynamicism share the view of it being an *alternative* to connectionism and computationalism broadly. (Hence the label 'extreme dynamicism' is used above to denote a narrower subclass of dynamicism.) Beer, for one, denies the extreme dynamicist thesis that cognitive systems are best understood *only* using the tools of dynamical systems theory (2013). He asserts that there is no useful mathematical distinction to be drawn among dynamicism, connectionism and (digital) computationalism. For, on the one hand, all dynamical systems can be approximated by TMs and, on the other hand, TMs defined over the real numbers are equivalent to dynamical systems. Similarly, recurrent connectionist networks can approximate arbitrary dynamical systems. Beer also acknowledges that it is probable that connectionism, (digital) computationalism and dynamicism will all be important in any future theory of cognition.

Nonetheless, mathematical distinctions aside, the mechanistic challenge remains unanswered. Connectionism and digital computationalism also make available models of cognitive architecture besides the mathematical toolbox that comes with the theory, but dynamicism proper does not. Digital computationalism need not be limited only to a specific formalism of computability, such as TMs,

Post machines or the lambda calculus. As already discussed in Chapter 1, formalisms of computability provide the mathematical tools required for determining the plausibility of *computational level theories*. Still, any particular formalism does not specify the relationship between *abstract* and *concrete* computation. And if cognition is an embodied biological phenomenon, as granted by the dynamicist, it is *concrete* computation that plays a key role in explaining cognition and not just *computability theory*.

As observed above, dynamicism and mechanistic computational explanations are complementary. Understanding a particular mechanism but not its role in the overall dynamics of the cognitive agent, and possibly the environment, is insufficient. Identifying a clock mechanism, for instance, in a physical computing system without discovering how it affects and is affected by the overall operation of the system only provides a partial explanation. And conversely, understanding the dynamics of the cognitive agents without identifying their constituent components provides a limited explanation at best.

This complementarity principle has yielded some collaborative effort in computational neuroscience where an understanding of single brain neurones is supplemented by dynamicist-type explanations. For instance, Eugene Izhikevich has applied dynamical systems theory tools in studying the relationship among electrophysiology, bifurcations and computational properties of neurones (2007). Chris Eliasmith and Charles Anderson have introduced a framework for the study of cognition in which computation, representation and dynamical systems theory – all play a role (2003). They argue that modern control theory is better suited than computability theory for understanding cognition as a biological system. According to their theory, neural computation is the transformation of neural representations.

It certainly seems plausible that cognitive science has much to gain by adopting a broad perspective, which sees the above paradigms as complementary. A bottom up strategy alone is likely to face significant challenges trying to explain how low level mechanisms give rise to high-level cognitive phenomena. A purely top down strategy may yield a viable story that explains certain phenomena without establishing how they are grounded in the human biological substrate. But such a story is difficult to empirically refute. An integrative cognitive science that draws on each of these strategies simultaneously is more likely to overcome those challenges. Time will tell.

8.4 Computation and Cognition – Concluding Remarks

As we have seen, the accounts of concrete digital computation examined in this book are non-equivalent. They are not just intensionally different, but also extensionally different. The PSS account denotes the narrowest class amongst the extensional classes of digital computing systems denoted by the accounts examined here. That is, the class of programmable stored-program digital computers. The mechanistic account and the IIP account denote the broadest class of digital computing systems without leading to digital pancomputationalism, which follows from Searle and Putnam's trivialisation of digital computation.

Since 'digital computation' is an ambiguous concept, any computational explanation of cognition is unintelligible without a commitment to a single interpretation of 'digital computation' as defined by a given account thereof.

As it turns out, most of the extant accounts of digital computation are inadequate when evaluated against the recommended set of adequacy criteria, which were introduced in Chapter 2. Only the mechanistic account and the IIP account are evaluated positively against these criteria. They are also preferable to semantic accounts of digital computation, for they do not posit any extrinsic representations. The IIP account can arguably be advantageous in the cognitive science discourse. For it accords with the common view that cognition is an information processing system and at the same time it shows that *computation* proper can be understood in terms of only *instructional* information, whereas natural cognition also inherently processes *declarative* information.

To evaluate the theses that natural cognition is an information processing system or that it is computational requires understanding what 'natural cognition' means. Progress in cognitive science is often made without an agreed upon definition of 'natural cognition'. For example, we can define cognition broadly as an evolutionary adaptable system that has an integral mechanism enabling the system to learn from its experience through an interaction with the environment. We can also assume, for simplicity, that cognition can be investigated independently of phenomenal consciousness, which involves the qualitative feel of conscious states. But, of course, the findings of a scientific research that is conducted based on some fundamental principles, working assumptions and paradigmatic examples of cognitive organisms are bound to have a limited applicability. As in the case of concrete computation, natural cognition too is in need of clarification. Only that an agreed upon definition of natural cognition does not seem forthcoming anytime soon, given the history of the notorious mind-body problem.

Nevertheless, it seems clear that natural cognitive agents identify states of themselves and their environment and use that information in conjunction with other stored information to guide the selection of their appropriate behaviour. They generally produce new information to deal with a variety of environments. If this new information is effective in guiding the agent's behaviour, it is likely to be retained. Otherwise, it is likely to be either modified or discarded to minimise overload on the agent's limited resources. Information-producing agents, thus, face a trade-off between enhancing their decision-making and being overloaded and slow in performing real-world, time-critical tasks. At the knowledge level, it seems that at least two distinct types of information are crucial for the cognitive agent: prescriptive and declarative. The former is required for knowledge-how and the latter for knowledge-that.

Research attention should be directed toward gaining a better understanding of the types of information processed by natural cognitive agents, how they are processed and interact and how such processing throws light on human cognitive architectures. Such research should examine how cognitive agents produce, acquire, extract, analyse and use information in learning, planning and decision-making, for example. It can inform contemporary cognitive science by identifying the mechanisms in the human cognitive architecture that are necessary for these information-processing operations. It seems clear that many existing models of cognitive agents, such as Belief-Desire-Intention models (cf. Rao, 1996; Torres, Nedel, & Bordini, 2003) and Belief-Change models of knowledge (cf. Alchourrón et al., 1985; Schurz, 2010), already make implicit use of different types of cognitive information.

But does the thesis that natural cognition is an information-processing system entail that natural cognition is simply a digital computing system? If the IIP account is correct, then the answer is negative. (Similarly, on the mechanistic account, cognition is a form of computation in a *generic*, rather than a purely *digital*, sense; (Piccinini & Scarantino, 2011, p. 34).) On the IIP account, digital computation can be understood solely in terms of *data* processing and *instructional* (or prescriptive) information processing, whereas natural cognition also processes *declarative* information. Natural cognitive agents adapt to unfamiliar environments by processing information about the current and future states of affairs. They make decisions based on the probability of this declarative information. Some of this information is also hypothetical: they assume and hypothesise about possible events based on past evidence. However, it seems implausible that a purely digital computing process suffices to produce the same cognitive behaviour. It is also interesting to point out that despite Turing's famous idea of the imitation game as a substitute for the attribution of intelligence to machines, he did not believe that cognition is purely a digital computing system – or a discrete state machine.

> "The nervous system is certainly not a discrete-state machine" (Turing, 1950, p. 451).

> "[Even if the] whole mind [were] mechanical [...it] would not be a discrete-state machine" (Turing, 1950, p. 455).

It is certainly plausible that some aspects of cognition are computational, whereas others cannot be simply reduced to computations. For instance, formal reasoning, which is based on a purely formal dynamic of signs that is independent from meaning, is what constitutes the TM moving along signs imprinted on its tape (Longo, 2009, pp. 48–49) and can be considered computational. Nevertheless, it need not apply to other complex and non-discrete aspects of cognition, such as sensation or imagination, that seem to be rich in causal circularities. Moreover, computational imitation, of the type proposed by Turing, makes deterministic chaos disappear completely by the discrete character of its data types. Natural cognition as a physical and biological phenomenon is extremely sensitive to the conditions of the environment, to the slightest variation in the parameters at play. Whilst a computational imitation game can deceive an observer of the game, it cannot model natural cognition (Longo, 2009, pp. 51–52), which resides in a biologically contingent brain.

Cognitive science faces the nontrivial task of explaining cognition. A blanket dismissal of the key role computation plays in cognitive science is unwarranted. Even if classicism, for example, were found untenable, digital computationalism could still survive. And indeed, as we have seen, classicism relies on a semantic view of computation that *is* untenable. Computation need not be representational, in the extrinsic sense, whereas cognition seems to be so. Although cognition as a whole does not seem reducible to digital computation, it seems clear that some cognitive functions *are* computational. It remains to be seen which cognitive aspects *are* computational and which are *not*. An integrative cognitive science is better geared to study how different cognitive processes interact and how different cognitive tasks are unified by a particular cognitive goal.

References

Alchourrón, C.E., Gärdenfors, P., Makinson, D.: On the Logic of Theory Change: Partial Meet Contraction and Revision Functions. The Journal of Symbolic Logic 50(2), 510–530 (1985), doi:10.2307/2274239

Anderson, J.R.: The architecture of cognition. Harvard University Press, Cambridge (1983)

Bechtel, W.: Representations and cognitive explanations: Assessing the dynamicist's challenge in cognitive science. Cognitive Science 22(3), 295–317 (1998a), doi:10.1016/S0364-0213(99)80042-1

Bechtel, W.: Dynamicists versus computationalists: Whither mechanists? Behavioral and Brain Sciences 21(5), 629 (1998b)

Bechtel, W.: Constructing a Philosophy of Science of Cognitive Science. Topics in Cognitive Science 1(3), 548–569 (2009), doi:10.1111/j.1756-8765.2009.01039.x

Bechtel, W., Abrahamsen, A.A.: Connectionism and the mind: parallel processing, dynamics and evolution in networks. Blackwell, Oxford (2002)

Bechtel, W., Richardson, R.C.: Discovering complexity: decomposition and localization as strategies in scientific research, 2nd edn. MIT Press, Cambridge (2010)

Beer, R.: Dynamical systems and embedded cognition. In: Frankish, K., Ramsey, W. (eds.) The Cambridge Handbook of Artificial Intelligence. Cambridge University Press, Cambridge (2013)

Boden, M.A.: Information, Computation, and Cognitive Science. In: Adriaans, P., van Benthem, J. (eds.) Philosophy of Information, vol. 8, pp. 741–761 (2008)

Brooks, R.A.: Intelligence without representation. Artificial Intelligence 47(1-3), 139–159 (1991), doi:10.1016/0004-3702(91)90053-M

Chalmers, D.J.: Subsymbolic computation and the Chinese room. In: Dinsmore, J. (ed.) The Symbolic and Connectionist Paradigms: Closing the Gap, pp. 25–47. L. Erlbaum Associates, Hillsdale (1992)

Churchland, P.S., Koch, C., Sejnowski, T.J.: What is computational neuroscience? In: Schwartz, E.L. (ed.) Computational Neuroscience, pp. 46–55. MIT Press, Cambridge (1990)

Churchland, P.S., Sejnowski, T.J.: The computational brain. MIT Press, Cambridge (1992)

Colburn, T.: Software, Abstraction, and Ontology. The Monist 82(1), 3–19 (1999), doi:10.2307/27903620

Craver, C., Bechtel, W.: Mechanism. In: Sarkar, S., Pfeifer, J. (eds.) The Philosophy of Science: An Encyclopedia, pp. 469–478. Routledge, New York (2006)

Dayan, P., Abbott, L.F.: Theoretical neuroscience: computational and mathematical modeling of neural systems. MIT Press, Cambridge (2001)

Diederich, J.: Spreading Activation and Connectionist Models for Natural Language Processing. Theoretical Linguistics 16(1), 25–64 (1990), doi:10.1515/thli.1990.16.1.25

Dodig-Crnkovic, G., Müller, V.C.: A Dialogue Concerning Two World Systems: Info-computational Vs. Mechanistic. In: Dodig-Crnkovic, G., Burgin, M. (eds.) Information and Computation, pp. 149–184. World Scientific (2011)

Eliasmith, C.: Computational neuroscience. In: Thagard, P. (ed.) Philosophy of Psychology and Cognitive Science, pp. 313–338. Elsevier (2007)

Eliasmith, C., Anderson, C.H.: Neural engineering: computation, representation, and dynamics in neurobiological systems. MIT Press, Cambridge (2003)

Fodor, J.A.: The language of thought. Harvard University Press, Cambridge (1975)

Fodor, J.A.: The Mind-Body Problem. Scientific American 244(1), 114–123 (1981), doi:10.1038/scientificamerican0181-114

Fodor, J.A., Pylyshyn, Z.W.: Connectionism and cognitive architecture: A critical analysis. Cognition 28(1-2), 3–71 (1988), doi:10.1016/0010-0277(88)90031-5

Fresco, N.: A Computational Account of Connectionist Networks. Recent Patents on Computer Science 3(1), 20–27 (2010), doi:10.2174/1874479611003010020

Froese, T.: Breathing new life into cognitive science. Avant 2, 113 (2011)

Hamann, H., Wörn, H.: Embodied Computation. Parallel Processing Letters 17(03), 287–298 (2007), doi:10.1142/S0129626407003022

Haugeland, J.: Artificial intelligence: the very idea. MIT Press, Cambridge (1985)

Hempel, C.G.: On the nature of mathematical truth. In: Fetzer, J.H. (ed.) Science, Explanation, and Rationality Aspects of the Philosophy of Carl G. Hempel, pp. 3–17. Oxford University Press, New York (2000)

Izhikevich, E.M.: Dynamical systems in neuroscience: the geometry of excitability and bursting. MIT Press, Cambridge (2007)

Kaplan, D.M., Bechtel, W.: Dynamical Models: An Alternative or Complement to Mechanistic Explanations? Topics in Cognitive Science 3(2), 438–444 (2011), doi:10.1111/j.1756-8765.2011.01147.x

Kirsh, D.: When is information explicitly represented? In: Hanson, P.P. (ed.) Information, Language, and Cognition, pp. 340–365. Oxford University Press, New York (1991)

Kremer, S.C.: Spatio-temporal connectionist networks. In: Fishwick, P.A. (ed.) Handbook of Dynamic System Modeling. Chapman & Hall/CRC, Boca Raton (2007)

Longo, G.: Critique of computational reason in the natural sciences. In: Gelenbe, E., Kahane, J.-P. (eds.) Fundamental Concepts in Computer Science, vol. 3, pp. 43–70. Imperial College Press, London (2009)

Maass, W., Markram, H.: On the computational power of circuits of spiking neurons. Journal of Computer and System Sciences 69(4), 593–616 (2004), doi:10.1016/j.jcss.2004.04.001

Machamer, P.: Activities and Causation: The Metaphysics and Epistemology of Mechanisms. International Studies in the Philosophy of Science 18(1), 27–39 (2004), doi:10.1080/02698590412331289242

MacLennan, B.J.: Connectionist approaches. In: Smelser, N.J., Baltes, P.B. (eds.) International Encyclopedia of the Social & Behavioral Sciences, 1st edn., pp. 2568–2573. Elsevier, New York (2001)

MacLennan, B.J.: Natural computation and non-Turing models of computation. Theoretical Computer Science 317(1-3), 115–145 (2004), doi:10.1016/j.tcs.2003.12.008

Marr, D.: Vision: a computational investigation into the human representation and processing of visual information. W.H. Freeman, San Francisco (1982)

Newell, A.: Physical Symbol Systems. Cognitive Science 4(2), 135–183 (1980a), doi:10.1207/s15516709cog0402_2

Newell, A.: The Knowledge Level. AI Magazine 2(2), 1–20 (1980b)

Newell, A., Simon, H.A.: Human problem solving. Prentice-Hall (1972)

Newell, A., Simon, H.A.: Computer science as empirical inquiry: symbols and search. Communications of the ACM 19(3), 113–126 (1976), doi:10.1145/360018.360022

O'Brien, G.: Connectionism, analogicity and mental content. Acta Analytica 22, 111–131 (1999)

O'Brien, G., Opie, J.: How do connectionist networks compute? Cognitive Processing 7(1), 30–41 (2006), doi:10.1007/s10339-005-0017-7

Pfeifer, R., Scheier, C.: Understanding intelligence. MIT Press, Cambridge (1999)

Piccinini, G.: Computers. Pacific Philosophical Quarterly 89(1), 32–73 (2008a), doi:10.1111/j.1468-0114.2008.00309.x

Piccinini, G.: Some neural networks compute, others don't. Neural Networks 21(2-3), 311–321 (2008b), doi:10.1016/j.neunet.2007.12.010

Piccinini, G., Bahar, S.: Neural Computation and the Computational Theory of Cognition. Cognitive Science 37(3), 453–488 (2013), doi:10.1111/cogs.12012

Piccinini, G., Craver, C.: Integrating psychology and neuroscience: functional analyses as mechanism sketches. Synthese 183(3), 283–311 (2011), doi:10.1007/s11229-011-9898-4

Piccinini, G., Scarantino, A.: Information processing, computation, and cognition. Journal of Biological Physics 37(1), 1–38 (2011)

Poggio, T., Koch, C.: Ill-Posed Problems in Early Vision: From Computational Theory to Analogue Networks. Proceedings of the Royal Society B: Biological Sciences 226(1244), 303–323 (1985), doi:10.1098/rspb.1985.0097

Pylyshyn, Z.W.: Computation and cognition: toward a foundation for cognitive science. The MIT Press, Cambridge (1984)

Pylyshyn, Z.W.: Computers and the symbolization of knowledge. In: Morelli, R. (ed.) Minds, Brains, and Computers: Perspectives in Cognitive Science and Artificial Intelligence. Ablex Pub. Corp., Norwood (1993)

Pylyshyn, Z.W.: What's in Your Mind? In: LePore, E., Pylyshyn, Z.W. (eds.) What is Cognitive Science?, pp. 1–25. Blackwell, Malden (1999)

Rao, A.S.: AgentSpeak(L): BDI agents speak out in a logical computable language, pp. 42–55. Springer-Verlag New York, Inc., Secaucus (1996)

Rubel, L.A.: The brain as an analog computer. Journal of Theoretical Neurobiology 4, 73–81 (1985)

Rumelhart, D.E., McClelland, J.L.: On learning the past tenses of English verbs. In: David, E., Rumelhart, J., McClelland, L. & C. PDP Research Group (eds.) Parallel Distributed Processing: Explorations in the Microstructure of Cognition, vol. 2, pp. 216–271. MIT Press, Cambridge (1986)

Schurz, G.: Abductive Belief Revision in Science. In: Olsson, E.J., Enqvist, S. (eds.) Belief Revision meets Philosophy of Science, pp. 77–104. Springer, Netherlands (2010)

Siegelmann, H.T.: Neural networks and analog computation: beyond the turing limit. Birkhauser, Boston (1999)

Siu, K.-Y., Roychowdhury, V.P., Kailath, T.: Discrete neural computation: a theoretical foundation. Prentice Hall PTR, Englewood Cliffs (1995)

Smolensky, P.: On the proper treatment of connectionism. Behavioral and Brain Sciences 11(01), 1–23 (1988), doi:10.1017/S0140525X00052432

Smolensky, P., Legendre, G.: The harmonic mind: from neural computation to optimality-theoretic grammar. MIT Press, Cambridge (2006)

Spivey, M.: The continuity of mind. Oxford University Press, Oxford (2007)

Stepp, N., Chemero, A., Turvey, M.T.: Philosophy for the Rest of Cognitive Science. Topics in Cognitive Science 3(2), 425–437 (2011), doi:10.1111/j.1756-8765.2011.01143.x

Thagard, P.: Mind: introduction to cognitive science, 2nd edn. MIT Press, Cambridge (2005)

Thelen, E., Smith, L.B.: A dynamic systems approach to the development of cognition and action. MIT Press, Cambridge (1994)

Thompson, E.: Mind in life: biology, phenomenology, and the sciences of mind. Belknap Press of Harvard University Press, Cambridge (2007)

Torres, J.A., Nedel, L.P., Bordini, R.H.: Using the BDI Architecture to Produce Autonomous Characters in Virtual Worlds. In: Rist, T., Aylett, R.S., Ballin, D., Rickel, J. (eds.) IVA 2003. LNCS (LNAI), vol. 2792, pp. 197–201. Springer, Heidelberg (2003)

Trappenberg, T.P.: Fundamentals of computational neuroscience, 2nd edn. Oxford University Press, Oxford (2010)

Turing, A.M.: Computing Machinery and Intelligence. Mind 59(236), 433–460 (1950)

Van Gelder, T.: The dynamical hypothesis in cognitive science. Behavioral and Brain Sciences 21(05), 615–628 (1998)

Van Gelder, T., Port, R.F.: It's about time: an overview of the dynamical approach to cognition. In: van Gelder, T., Port, R.F. (eds.) Mind as Motion: Explorations in the Dynamics of Cognition, pp. 1–44. MIT Press, Cambridge (1995)

Van Rooij, I.: The tractable cognition thesis. Cognitive Science 32(6), 939–984 (2008), doi:10.1080/03640210801897856

Varela, F.J., Thompson, E., Rosch, E.: The Embodied mind: cognitive science and human experience. MIT Press, Cambridge (1991)

Wallace, B., Ross, A., Davies, J., Anderson, T. (eds.): The mind, the body, and the world: psychology after cognitivism? Imprint Academic, Exeter (2007)

Weiskopf, D.A.: Models and mechanisms in psychological explanation. Synthese 183(3), 313–338 (2011), doi:10.1007/s11229-011-9958-9

White, G.: The Philosophy of Computer Languages. In: Floridi, L. (ed.) The Blackwell Guide to the Philosophy of Computing and Information, pp. 237–247. Blackwell, Malden (2004)

Zednik, C.: The Nature of Dynamical Explanation. Philosophy of Science 78(2), 238–263 (2011), doi:10.1086/659221

Printed in the United States
By Bookmasters